TRACE EVIDENCE ANALYSIS

MORE CASES IN MUTE WITNESSES

Edited by Max M. Houck
Director, Forensic Science Initiative
West Virginia University
Morgantown, WV

Executive Director
Institute for Cold Case Evaluations, Corp. (ICCE)
Morgantown, WV

ELSEVIER
ACADEMIC
PRESS

Amsterdam Boston London New York Oxford Paris
San Diego San Francisco Singapore Sydney Tokyo

This book is printed on acid-free paper.

Copyright © 2004, Elsevier (USA). All rights reserved.

No part of this publication may be reproduced, stored in a retrieval system, or transmitted in any form or by any means electronic, mechanical, photocopying, recording or otherwise, without the prior written permission of the publisher.

Permissions may be sought directly from Elsevier's Science & Technology Rights Department in Oxford, UK; phone: (+44) 1865 843830, fax: (+44) 1865 853333, e-mail: permissions@elsevier.co.uk. You may also complete your request on-line via the Elsevier homepage (http://www.elsevier.com), by selecting 'Customer Support' and then 'Obtaining Permissions'.

Elsevier Academic Press
200 Wheeler Road, 6th Floor, Burlington, MA 01803, USA
http://www.elsevier.com

Elsevier Academic Press
84 Theobald's Road, London WC1X 8RR, UK
http://www.elsevier.com

Library of Congress Catalog Number:

British Library Cataloguing in Publication Data
A catalogue record for this book is available from the British Library.

ISBN 0-12-356761-0

Printed and bound in Italy
04 05 06 07 08 9 8 7 6 5 4 3 2 1

CONTENTS

ACKNOWLEDGEMENTS		vii
CONTRIBUTORS		ix
	INTRODUCTION Michael Grieve and Max M. Houck	1
CHAPTER 1	HAIR OF THE DOG: A CASE STUDY Silvana R. Tridico	27
2	FIBER-PLASTIC FUSIONS AND RELATED TRACE MATERIAL IN TRAFFIC ACCIDENT INVESTIGATION Georg Jochem	53
3	AN INTELLIGENCE LED INVESTIGATION USING TRACE EVIDENCE Ray Palmer	89
4	THE VALUE OF SOIL EVIDENCE Thomas J. Hopen	105
5	THE IMPORTANCE OF TRACE EVIDENCE Harold Deadman	123
6	CEREAL MURDER IN SPOKANE William M. Schneck	165
7	USING 1:1 TAPING TO RECONSTRUCT A SOURCE Kornelia Nehse	191
8	WHO DO YOU BELIEVE? Barbara P. Wheeler	211
9	MY ROOMMATE IS USING THE REFRIGERATOR Max M. Houck	233
AUTHOR INDEX		251
SUBJECT INDEX		253

The year 2002 took a horrible toll on the field of trace evidence analysis. The loss to forensic science with the passing of Jim Crocker, Walter McCrone, and Mike Grieve can hardly be accounted. Having known and learned from these men is not enough; we must, at every chance, continue the work to which they passionately dedicated their lives. This book is dedicated to their living memory.

This book is also dedicated to Samia, Eric, and Ben. Despite what others may have thought, you were just being yourselves and deserved better.

ACKNOWLEDGEMENTS

The authors would like to thank their respective agencies for the time and resources to work on this project.

I would like to thank my colleagues, friends, and my father for their encouragement throughout my life and career. I would also like to thank Lucy for her support, tolerance, and love, not necessarily in that order.

The authors of *Trace Evidence Analysis: More Cases in Mute Witnesses* also deserve acknowledgement. I was overjoyed at finding more experts willing to extol the values of trace evidence. Recommend this book to every attorney, judge, laboratory director, and law enforcement officer you know.

Thanks, also, to Mark Listewnik and the professionals at Academic Press for making this book happen...again.

Finally, I'd like to acknowledge the man I met from the Pentagon whose life and work were cut short while drinking coffee at his desk on the morning of September 11, 2001. I think about him every morning as I look out my office window and sip my coffee.

The proceeds from this book go to the Max W and Janet P. Houck Forensic Science Endowment at Michigan State University, East Lansing, Michigan. The grant is competitive for graduate students pursuing their degrees in forensic science and/or forensic anthropology at MSU.

In assaults, murders, rapes, and kindred crimes, contact between the criminal and his victim is the rule, and always leads to the interchange of fibers, hairs, ducts, and fragments of microscopic dimensions. In addition to clothing, the finger nail scrapings, ear wax, shoe soles, and other likely places accumulate traces of similar materials. It is the belief of the author, as a result of years of study of this type of evidence in many varieties of crime, that only a minute percentage of such evidence is ever exploited even to a reasonable extent, and that the majority of unsolved crimes are those in which this type of evidence was entirely or partially neglected.

Paul L. Kirk (1949) "Microscopic evidence – its use in the investigation of crime", *Journal of Criminal law, Criminology, and Police Science*, Vol. 40, pp. 362–369. (Quote taken from pp. 363–364.)

CONTRIBUTORS

Max M. Houck (Editor)
Director, Forensic Science Initiative, West Virginia University, 886 Chestnut Ridge Road, PO Box 6216, Morgantown, WV 26506.

Executive Director, Institute for Cold Case Evaluations, Corp. (ICCE), Morgantown, WV

Michael Grieve
Forensic Science Institute, German Federal Police Office, Wiesbaden, Germany.

Silvana R. Tridico
Forensic Scientist, Centre of Forensic Sciences, Toronto, Canada.

Georg Jochem
Forensic Science Institute, Fiber section, German Federal Police Office, Wiesbaden, Germany.

Ray Palmer
The Forensic Science Service, Huntingdon, UK.

Thomas J. Hopen
Bureau of Alcohol, Tobacco, Firearms and Explosives, Forensic Science Laboratory, Atlanta; Arson and Explosives Section, Atlanta, GA, USA.

Harold Deadman
George Washington University, Department of Forensic Science, 2036 "H" Street, Samson Hall, Washington, DC 20052, USA.

William M. Schneck
Microanalysis Section, Washington State Patrol Crime Laboratory, West 1100 Mallon, Public Safety Building, Spokane, Washington, WA 99210, USA.

Kornelia Nehse
Textile Expert and Head of the Fibers Group, German State Police, Karl-Marx-St. 14, D-14612 Falkensee, Berlin, Germany.

Barbara P. Wheeler
Forensic Scientist, Lexington, KY.
(Former Forensic Laboratory Supervisor, Trace Evidence Unit, Kentucky State Police Forensic Laboratory Section).

INTRODUCTION

Michael Grieve
Forensic Science Institute, German Federal Police Office, Wiesbaden, Germany

Max M. Houck
*Forensic Science Initiative, West Virginia University
Executive Director, Institute for Cold Case Evaluations, Corp. (ICCE)
Morgantown, WV*

Trace Evidence Analysis: More Cases in Mute Witnesses is a continuation of forensic case studies highlighting the use of trace evidence. The phrase "trace evidence" can be defined as microscopic material recovered as evidence that is used to help solve criminal cases. Because of their minute nature, trace materials can be easily cross-transferred from one surface or substrate to another without detection by a criminal.

French scientist Edmund Locard (1877–1966) (Locard, 1930) believed that every criminal could be linked to the crime she or he committed by the examination of transferred trace materials. He postulated his famous "exchange principle" which has been characterized as "every contact leaves a trace". The historical details about Locard have been covered in *Mute Witnesses* (Houck, 2001), but a simple analogy will demonstrate the exchange principle.

A trace evidence analyst may be defined as a forensic scientist concerned with the characterization, identification, and comparison of microscopic materials in criminal cases. This definition is deliberately all-encompassing because of the infinite variety of materials present on the earth of which trace materials comprise a tiny subset. Trace evidence helps solve crimes by linking people, places, and things involved in a crime by the microscopic materials they share through contact. The presence of indistinguishable trace materials on items from different sources demonstrates that they may have come into contact. In a legal sense, these traces only become evidence after they are introduced into the courtroom as exhibits. The value or significance of trace material findings is dependent upon the type(s) and amount(s) of trace evidence, the location(s) where they are found, and the circumstances of the crime.

In addition to providing associative evidence, evaluation of trace evidence recovered at a crime scene at the onset of an investigation may provide investigative information which may prove helpful to the police in their attempts to develop a suspect. This applies particularly to transferred fibers. In addition,

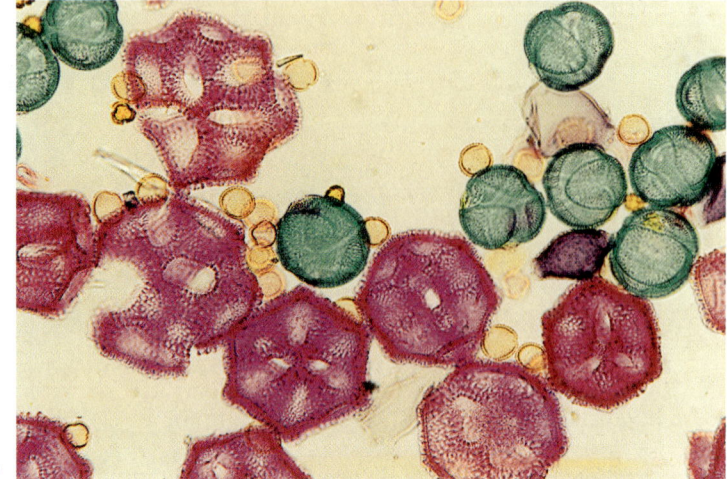

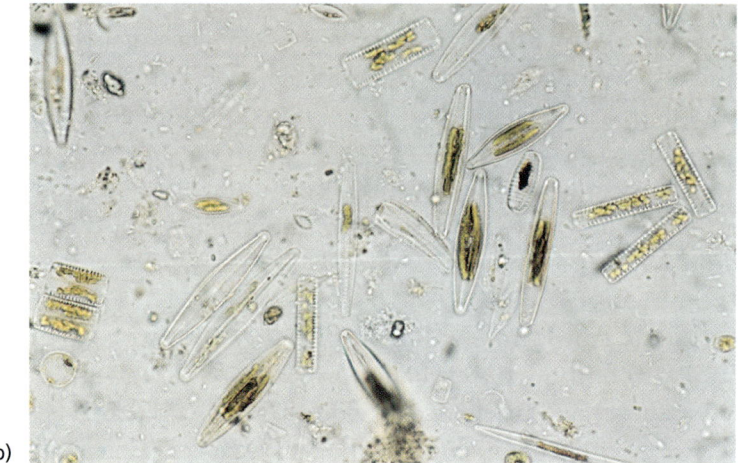

Figure I.1
(a) Arizona pollen grains (photo courtesy of Michael Eyring, Arizona DPS).
(b) Examples of different varieties of Diatoms (photo courtesy of Helmut Demmelmeyer, Bundeskriminalamt, Germany).

the epitome of localized evidence collection is the appropriately-named technique of "1:1 Taping," where the area covered by one piece of tape represents the same area of the surface taped, can help reconstruct events that may have taken place during the crime. However, this type of evaluation is exceedingly complex and, therefore, is best dealt with only by experienced personnel trained in this method.

Although the most frequently encountered forms of trace evidence are human hairs, fibers, paint, and glass, trace cases may require examination of a wide variety of microscopic detritus, such as plant material, wood, seeds, fungal spores, pollen grains (see Figure I.1a), insects, animal hairs, soil, sand, mineral particles, diatoms (see Figure I.1b), feathers, metal particles, gunshot residues, explosives, adhesives, plastics, lubricants, and cosmetics. The list is virtually never-ending; *anything* can become evidence. Cosmetics may seem like an odd example, but the transfer of lipstick, makeup, glitter particles, and broken pieces of polished fingernail have all played a part in forensic cases. Although not usually thought of as trace evidence, body fluids and excretions can be defined as transfer evidence, or, when the quantities are small enough, trace evidence. Unlike most other forms of trace evidence these will remain in their original place of deposition and may therefore play an important interpretive role.

Trace materials will characterize a particular environment and, even if individually they are not unique to it, the chances that the same combination of "ingredients" will be found together elsewhere becomes infinitesimally small. This is especially true as the number and type of ingredients increases.

For example, the textile population in a person's wardrobe and living area is based upon personal taste, fashion, economics, and climate, among other factors, and is therefore highly individual. Foreign fibers picked up on a person's clothing from his or her living environment are also fibers that may be secondarily transferred in a crime. Consider how extremely unusual it would be to find two individuals wearing *exactly* the same clothing from top to toe (even when only comparing color subjectively) among a large crowd at a concert or sporting event. How much more unusual would it be to find a shared population of indistinguishable foreign fibers on suspect and victim items when the fibers can be subjected to highly discriminating analytical examinations?

As stated previously, particles of any substance may provide trace evidence indicating that a transfer has occurred. The usual aim of searching for transferred traces is to try to provide evidence of contact between persons and/or objects. These links or associations are used to corroborate witness statements and rebut denials by the accused that he was in a particular location or engaged in a particular activity. Not finding traces on certain items may be just as significant: *It is important to remember that a negative finding does not automatically mean that no contact took place.* This has been phrased informally as, "Absence of evidence is not evidence of absence." It may be possible to offer explanations for this "lack of evidence." For example, because trace evidence is lost at a geometric rate, the elapsed time interval between the crime and seizure of evidence may have been such that any transferred trace evidence has fallen off. Or even though a textile contacted another substrate, it may have had a limited shedding potential, for example, a windbreaker constructed from filament yarn, giving it a smooth shiny surface. The findings of an examination, whether positive or negative, may be used to help reconstruct the crime and counter fabricated alibis.

THE INCREASING COMPLEXITY OF TRACE EVIDENCE CASES

One of the disadvantages of trace evidence, contrary to its portrayal in films and by the media, is that the laboratory work is labor-intensive and often takes

Figure I.2
The Maxcan fiber finder system from Cox Analytical Systems AB, Gothenburg, Sweden (reproduced with permission of "Science & Justice").

[1] For example, in 2000, my technician and I processed over 2000 items of evidence in 1 year of trace evidence casework (MMH).

a long time. A complex fiber case may involve samples from several persons and locations. There may be hundreds of bulk items to examine including clothing, bedding, and other household textiles, plus trace tapings or sweepings containing many thousands of individual fibers.[1] All of these, or samples from them, have to be painstakingly examined under the microscope. A result is more likely to be expected in terms of weeks or even months, rather than in hours or days. Microscopical examination of trace evidence is a task that requires a special ability to concentrate and enormous diligence and patience.

Attempts to automate the search and recovery of fibers and particles from trace tapings by the use of an automated fiber finder system (see Figure I.2) were greeted with something akin to euphoria when equipment from various companies became available about four years ago. The principle is that the instruments measure and store color data from reference samples and then search for suspect fibers and particles having matching color coordinates. The ability of the instruments to recognize some shapes requires additional research.

Due to both hard- and software limitations, the instruments, while being partially successful, have not yet achieved the expected breakthrough.

Not only can an enormous variety of different materials be encountered as trace evidence, but even within individual specialized categories, sample diversity has increased steadily over the years. Particularly good examples of this are manufactured fibers and paint. Reference to the classical text book on Forensic Medicine by Smith and Fiddes (1955) shows that in 1955, much greater emphasis tended to be placed on the transfer of hairs rather than fibers; the only types of fibers referred to were wool, cotton, and silk. About 30 different generic types of manufactured fibers are now in production and within some of these generic types there are many variations in polymer compositions resulting in sub-classes. In addition, manufactured fibers today exhibit an ever-increasing variety of cross-sectional shapes within one generic type. All of these different chemical and morphological differences are of great value to the forensic scientist, as they help individualize materials. In the past, fibers were mainly considered in connection with the possibility that suspect clothing fibers might be found under a victim's fingernails in murder cases. With the advent of taping as a collection technique for trace materials in Europe in the 1960s, and later in the US, fiber associations became possible in all types of cases.

The following illustrates the amount of detail that can be encountered in just one class of trace evidence: fibers. Fourier transform infrared microspectroscopy (micro-FTIR) is an analytical technique that can not only determine the chemical composition of a substance but, if the composition is very specific, it may help determine the manufacturer. In recent years micro-FTIR has made it possible to make much more detailed analyses of some generic types of fibers; about 40 different sub-classes of fibers containing acrylonitrile can now be identified. Micro-FTIR is routinely performed on very small fragments of single fibers, perhaps only $100\,\mu m$ long. With the old dispersive infrared spectrometers the amount of material required was much greater and sample preparation procedures were not only much more complicated, but differed according to the generic class of fiber analyzed.

In addition to the basic fiber polymer, components which can be detected by micro-FTIR include co-polymers which act as plasticizers, ter-polymers which are added to provide dye sites, chemicals added to confer flame retardancy, and residues of solvents used in the manufacturing process. Together with an examination of the fiber cross sectional shape (see Figure I.3), these features may provide information pointing toward a specific manufacturer. The frequency of occurrence of these characteristics among the acrylic fiber population will vary depending upon production levels, with the less frequently occurring types having more evidential value.

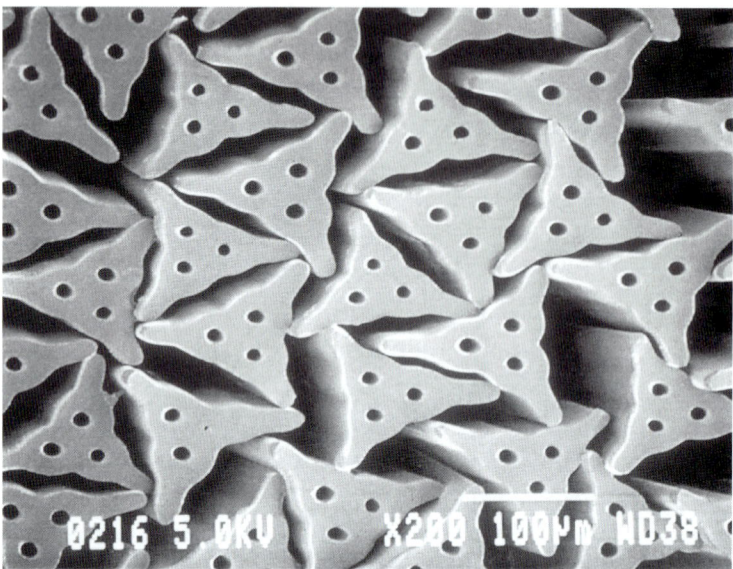

Figure I.3

"Nobel" hollow Nylon 6 carpet fibers (courtesy of Donna Knoop, Allied Signal fibers).

It is important for forensic scientists to keep abreast of new developments, especially in the automotive coatings and textile industries. Not only must fiber analysts be aware of new fiber types appearing on the market, but an understanding of textile technology may be of great value. This knowledge is especially beneficial in the following types of fiber examinations:

- establishing the cause of textile damage,
- reconstructing textile remains (e.g. after fire damage or long term immersion in water),
- garment sourcing, which may assist in the identification of a corpse,
- fabric impressions and fiber-plastic fusions which often result from traffic accidents (see Chapter 2), and
- examination of ropes and cordage (see Chapter 9).

The development of new fiber types is fueled by the search for new raw materials that are more economic and less damaging to the environment. Examples are Lyocell, a type of cellulosic fiber with a new environmentally-friendly production process; lactic acid fibers (now being marketed by Cargill Dow Polymers in the USA under the trade name "Nature Works"); and polybutylene terephthalate (PTT) where the raw material (1,3-propanediol) can be produced from corn starch. Polylactic acid was first produced in Japan by Kanebo in 1998 and the fibers, which have inherent low-flammability have found applications in apparel, non-wovens, carpeting, and furnishings. Like PTT, it is also produced from plant sugars and, despite being a melt-spun polymer, it is fully recyclable under composting conditions.

Other recent developments include production of anti-bacterial and anti-microbial fibers of different generic types (polyester, viscose, acrylic, and polypropylene) and channeled fibers that wick perspiration from the body, combating odor retention and leading to increased wearer comfort. Fiber cross sectional shape is a morphological property which influences physical and aesthetic properties of fibers; for example, a trilobal cross-section is used in carpet fibers to hide dirt and create a plush feel. Manufacturers are continually searching for new cross sectional forms and a truly remarkable range of the very latest shapes for bicomponent fibers can be seen, for example, in information on fiber extrusion equipment produced by the Hills, Inc. Company ⟨www.hillsinc.net.⟩

In the recent past, polypropylene (PP) has shown the greatest dynamism and growth of all generic groups of manufactured fibers. Polypropylene fibers have several very desirable properties but have hitherto suffered from the disadvantage that they could only be colored by incorporation of pigments into the molten polymer. Much research has been devoted to producing dyeable polypropylene.

The diversity of fiber types encountered in forensic cases today is considerable. Unusual examples from recent cases seen by one of the authors (MG) include polynosic fibers (a special form of cellulosic fiber made in Japan) left on the body of a murder victim; three different varieties of flame resistant Aramid fibers in an arson case requiring examination of fire-fighters' uniforms; split-film polypropylene fibers in cordage used to attach weights to a submerged corpse; ultra-fine nylon microfibers from a sports glove; and acrylic fibers containing ethylene carbonate solvent residue which originated from a blanket. In this last case, only two companies in the world use this solvent to produce acrylic fibers and the total amount of fiber produced annually is relatively very small. The forensic fiber expert must be able to identify unusual fiber types, and have some knowledge about fiber composition, properties, manufacturing processes, and end uses.

As mentioned earlier, sometimes it becomes necessary for the forensic scientist to try to trace the origin of a garment in order to help identify a victim. This may happen if it is suspected that the victim lived in a country other than where the body was discovered. Such enquiries may lead to a complicated chain of events involving border crossings. A textile may be manufactured in one country (from fibers produced elsewhere), assembled in a second, sold to an importer/distributor on a different continent from which it finally find its way to a retail outlet. Of course, most textiles are not unique, but batch numbers are often lower than might be expected (i.e. in the order of hundreds rather than thousands), and sometimes it is possible to research sales figures from various outlets. The ability to communicate quickly using e-mail and to use the search facilities available on the Internet has greatly simplified such enquiries.

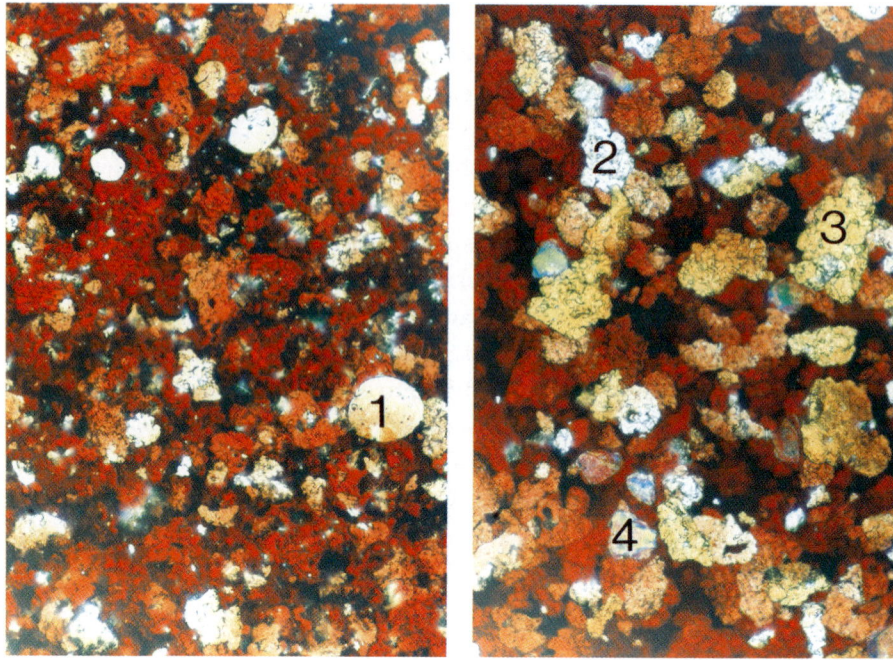

Figure I.4
Aluminum dollar (1) and cornflake (2) pigments in a paint flake seen against a background of red pigment particles. In addition golden cornflakes (3) coated with Fe_2O_3 can be seen, as can interference pigment particles (4) of mica coated with titanium dioxide (photo courtesy of Wilfried Stoecklein, Bundeskriminalamt, Germany).

In the area of paint technology, pigments can provide an example of increasing complexity. A recent development is the introduction of plate-like pigments to create clearer, more brilliant and exciting colors in the automotive paint market. Market research shows that car buyers prefer coatings with high chroma shades based on "effect" pigments because of their visual impact. The number of pigment types has increased considerably in the last 10 years and continues to increase.

These pigments can be of several different types. In addition to the normal "one dimensional" organic and inorganic pigments, "two and three dimensional" plate-like pigments are now finding their way into the market. The optical impression of these pigments is based on three different kinds of interactions with light. If the particles are small metallic flakes of over 5 μm diameter, they will act as small mirrors, causing direct reflection of light and producing lightness which "flops" with different viewing angles. Aluminum "dollar" and "corn flake" (see Figure I.4) pigments belong to this group. If the plate-like particles are based on or contain selectively absorbing materials, like plate-like iron oxides and plate-like phthalocyanine, the interaction with light is based on absorption *and* reflection, causing both reflected and transmitted light to be colored.

Three-dimensional plate-like pearlescent pigments are based on highly refracting materials, like bismuth oxychloride and titanium dioxide coated micas. These can be used to produce interference and reflection effects. The colors seem to flutter and change; strongly angle-dependent colors can be seen with various new luster pigments containing optically variable properties. There are two structurally different types of pigments with optically variable properties, the Fabry-Pérot type and the inner reflector type. Other new types of finishes are holographic pigments and interference pigments on the basis of liquid crystals.

EXAMINATION AND METHODOLOGY

The foundation for success in trace evidence cases, even more so than with other types of cases submitted to forensic laboratories, lies in efficient and timely collection of the evidence at the crime scene. Some important general criteria relating to the collection of trace evidence have been defined by Caddy (2000). Because of the microscopic nature of trace evidence, it can be quickly lost or re-transferred. This brings with it the additional crucial need to ensure that no form of accidental trace evidence contamination takes place after the crime. In recent years an increased investment in training crime scene personnel and strict quality assurance requirements in forensic science laboratories have minimized the possibility of such an occurrence.

Not only must crime scene personnel be thoroughly trained in appropriate methods of bulk evidence collection and packaging, but also they must give careful thought to the collection of "invisible" microscopic crime scene samples. The investigator should consider various possible scenarios, such as "if this had taken place, what may I expect to find in terms of trace evidence?" Trace tapings of nude bodies should be performed before the body is moved. The floor under and around a body should be taped to recover the most recently deposited trace materials. With an eye toward sheddability and unusual color, the investigator must assess which bulk items at a scene may have contributed trace materials and will, therefore, require sampling as standards.

The relative merits of microscopical versus instrumental analysis of trace evidence may be debated, focusing on such issues as sensitivity, discriminating power, sample size, sample destruction, etc. But one thing is certain: the most sophisticated analytical instrument in the world is of no help at all to a trace evidence analyst if the particles of interest cannot be *located*, *recognized* and *separated* before *analyzing* them. The first three steps require a microscope.[2] A great advantage of microscopy is that it enables the analyst to distinguish between polymorphs. Polymorphs are substances that have the same chemical composition but can crystallize with different internal structures, producing differences in external morphology and internal physical properties. Common examples are carbon, calcium carbonate, silica, and titanium dioxide.

During the 1890s Professor Emile Chamot of Cornell University claimed that there were few, if any, chemical problems that could not be solved or nearly

[2] This point is also made with hairs in Chapter 1.

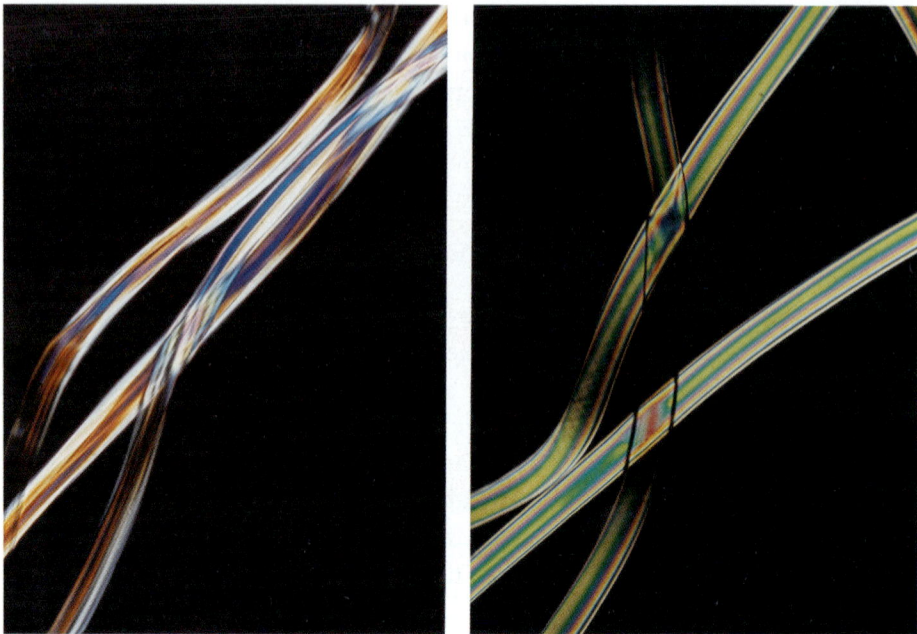

Figure I.5
Irregular viscose rayon fibers (left) and trilobal polypropylene fibers (right) viewed under polarized light/crossed polars.

solved by chemical microscopy. As his student, Dr. Walter McCrone, has noted, the application of chemical microscopy to forensic problems changes it into forensic microscopy. This includes the characterization, identification, and comparison of all types of trace evidence.

Many forensic scientists worldwide, but particularly in the United States, owe a tremendous debt of gratitude to Dr. McCrone. There can be very few trace evidence analysts who have not at sometime or other benefited from participation in one or more of the numerous microscopy courses available from the McCrone Research Institute in Chicago. They have also benefited, directly or indirectly, from Dr. McCrone's undying enthusiasm for microscopy, particularly polarized light microscopy (PLM). PLM is the basis for identification of synthetic fibers (see Figure I.5) and for mineralogical examinations; it is routinely used by analytical microscopists working in many different fields. Today, many other forms of microscopy are available to the forensic scientist: interference microscopy, dispersion staining microscopy, thermal microscopy, fluorescence microscopy, microspectrophotometry, and many others as technology advances. Even infrared spectroscopy is linked to microscopy as micro-FTIR; Figure I.6a and b shows examples of old and new comparison microscopes, another tool invaluable to the trace analyst.

Through his extensive knowledge, Dr. McCrone invented elegant solutions to complex problems, achieving the same results that would normally require much more expensive instrumentation. A good trace evidence analyst must use

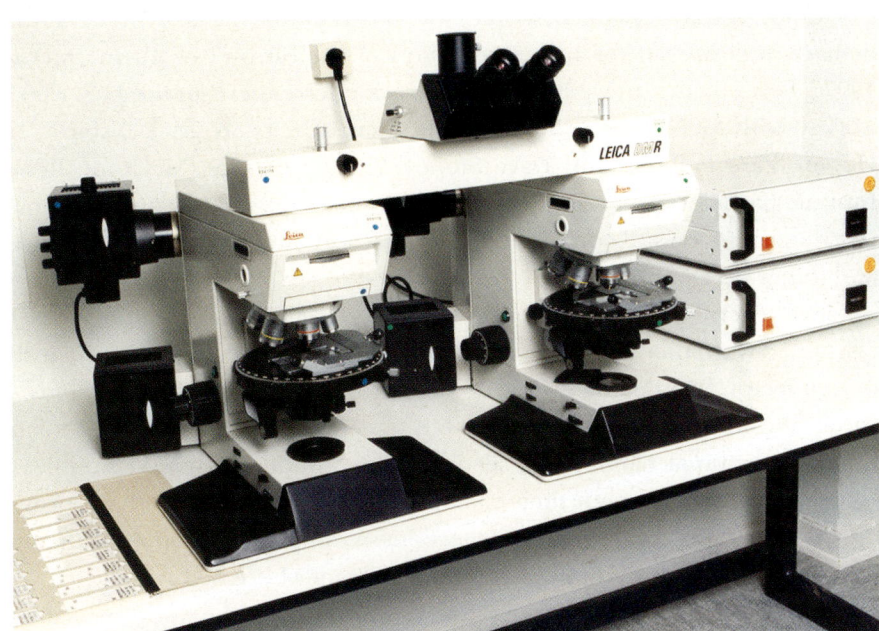

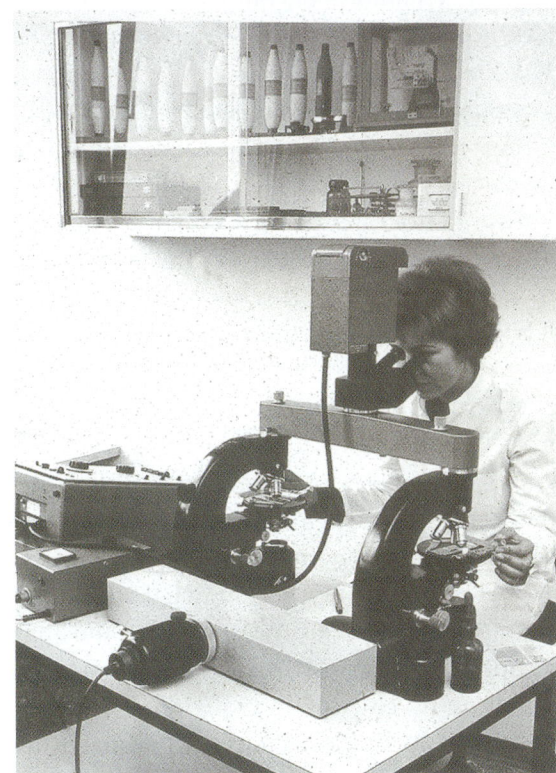

Figure I.6
Thirty-five years between (a) new (c. 2000) and (b) old (c. 1965) Leica comparison microscopes (reproduced with permission of "Science & Justice").

a flexible approach, as microscopic case materials may vary. Dr. McCrone is particularly well known for his examination of the Shroud of Turin (McCrone, 1997). Several current well-known trace evidence experts in the U.S. who have worked with McCrone at some stage of their careers are Dick Bisbing, Thom Hopen, Skip Palenik, and Dave Stoney. Since 1997, the McCrone Research Institute has offered a program for certification in applied Chemical Microscopy, based upon successful completion of at least six courses offered by the Institute and follow up examinations.

Most "legendary" trace evidence specialists share the fact that they have been fascinated with microscopy from an early age. It also helps to have the instincts of a collector. People in a variety of professions may collect stamps, coins, postcards, and memorabilia of all kinds but trace evidence analysts tend to collect fibers and animal hairs, sand, seeds, pollens, wood and paper samples, and spend time learning about their characteristics and origins. In other words, they are building their own reference collections. A sample may crop up as the subject of a case that the analyst knows he has seen somewhere before.

There are many capable forensic experts throughout the world specializing in the more commonly-encountered types of trace evidence. Several case examples have been included within the pages of *Trace Evidence Analysis: More Cases in Mute Witnesses* as well as the original *Mute Witnesses* book. But when it comes to particle identification, there is one name – Skip Palenik – that stands out. Palenik was a former colleague of Dr. Walter McCrone and learned to use a microscope when he was eight years old. He has since been involved in many, many high profile trace evidence cases that have occurred in the US in recent years.

Cases involving particle identification are much less frequently documented than those involving the more common forms of trace evidence and, therefore, it is appropriate to include a brief synopsis of two of Palenik's sand cases.

CASE 1

On taking delivery of crates containing a shipment of computers from Texas, the recipient in Argentina found that thieves had replaced them with concrete blocks. It was important to know where the switch had been made: before, during, or after the journey. On examining the samples of concrete, Palenik found that all the sand grains used as an aggregate in the mix were about the same size: characteristic of beach sand, where wave action sorts the sand grains by size. Under the scanning electron microscope, it was apparent that there were triangular gouges on the faces of the grains, probably due to them having been slammed into one another, which might occur in a coastal location.

Using a polarizing microscope to examine their optical properties, he was able to isolate and identify grains characteristic of the minerals zircon, staurolite, cyanite, and sillimanite. This is a mix most likely to have originated from eroded metamorphic rock. By perusal of the survey records of geologists employed by oil companies (who believed in the 1930–1940s that heavy minerals might be indicative of petroleum-rich strata), Palenik found a close match for his sand in southern Florida. Investigators knew that the computer shipment had passed through Miami International Airport on its way to Argentina and, to their amazement; they found identical concrete blocks on a construction site in a corner of the airport.

CASE 2

Another case involved a photographic company that sent spent developing solution to Colorado by rail, where the silver was extracted, placed in milk cans, and returned to the East Coast. One shipment of returned silver was stolen: the cans were full of sand. Where was the switch made? On washing the sand, Palenik looked for pollen grains. Pollen was present that was characteristic of sagebrush plants (*Artemesia*) that grow in the Great Plains and Rocky Mountains. Further examination of the sand revealed a particular form of zircon. A sample of sand given to Palenik from a friend, who had visited Colorado, also contained identical zircons. He then asked investigators to obtain samples from a train station in Colorado which lay *en route*, and obtained a perfect "match"! It is both fascinating and amazing that such precise information can be obtained from the examination of seemingly insignificant tiny particles.[3]

[3] See Chapter 4 for further examples of forensic soil examinations.

RECENT ADVANCES IN TRACE EVIDENCE ANALYSIS

A number of standard instrumental techniques can be applied to the analysis of more than one type of trace evidence; these include infrared microscopy, microspectrophotometry, and scanning electron microscopy in conjunction with elemental analysis, gas and high performance liquid chromatography and mass spectrometry. New analytical techniques, which are more sensitive, more discriminating, or quicker, are always under consideration.

In the field of fiber examination, the greatest analytical advance recently has been the introduction of diode array spectrophotometer systems (see Figure I.7). Because these are capable of scanning the whole spectral range simultaneously, they have greatly speeded up the acquisition of spectral data: A spectrum can be measured in less than a second. This has two great advantages. Firstly, a large number of suspect fibers can be scanned very quickly to establish if matching fibers or collectives are present. And secondly, the facility to rapidly generate and examine data from a large number of fibers makes research into fiber frequencies and fiber populations more feasible than ever before.

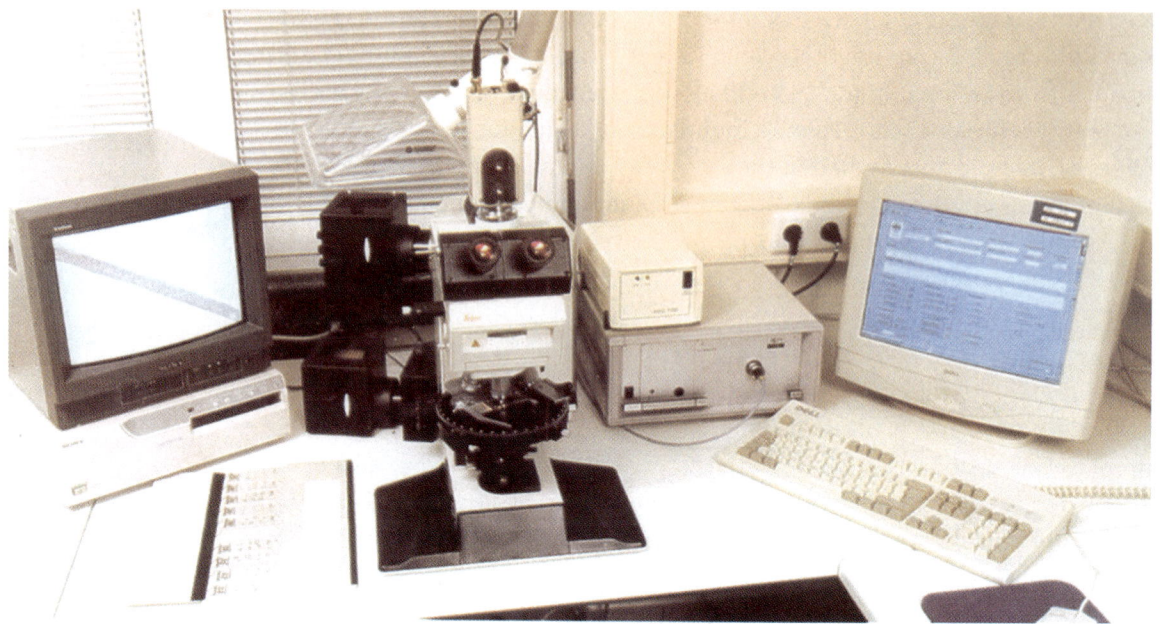

Figure I.7
Fiber work station incorporating J&M TIDAS Spectrophotometer system based on a Leica DMX microscope equipped for polarized light and fluorescence examinations.

A new method, which may find application in future fiber examinations, is the use of atomic force microscopy. The number of publications pertaining to this technique in connection with single fiber analysis is growing. It permits a detailed examination of surface morphology and thus allows observation of features such as microfibrils that may permit distinctions to be made between fibers of the same type originating from different manufacturers, or between those having been subjected to different production processes within the same company. The observed structures can be correlated with birefringence and fiber tenacity. This ability to progress further toward individualization of manufactured fibers would be of tremendous value in acquiring information about sourcing at the fiber, rather than the garment, level. At present, sourcing is very difficult because of the large number of fiber manufacturers worldwide that produce very similar products (e.g. round polyester fibers). The reader should not take this to mean that it is currently impossible to distinguish between round polyester fibers which have different origins – that is certainly not so. But the possibilities for determining their manufacturer, particularly with fibers that are undyed, can be relatively limited. As well as being useful for distinguishing between closely related manufactured fibers (cellulosics and olefins), atomic force microscopy can also differentiate between sheep's wool and various species of specialty animal hairs by accurate measurements of scale patterns.

Raman microspectroscopy has generated considerable interest lately in connection with fiber examination. This technique is closely related to micro-FTIR but in contrast to it, non-polar entities (e.g. unsaturated or aromatic compounds) such as organic dyestuffs tend to give stronger signals than the polar molecular structures comprising the fiber polymer. Because of this, the technique is likely to remain inferior to micro-FTIR as a means of identifying manufactured fibers but offers the potential of being able to identify fiber colorants. Dye peaks may be seen in infrared spectra, but at best are only able to provide some general information on the type of dye involved. Microspectrophotometry, while providing a non-destructive technique for color comparisons on minimal quantities of single fiber, does not (normally) permit the colorant to be identified.

There are some problems with Raman microspectroscopy. For example, short wavelength laser sources tend to induce fluorescence in samples, which may swamp the signal. Also, an extensive databank of dye spectra would have to be developed to permit dye and/or pigment identification. Raman may prove useful in the examination of producer-colored pigmented fibers that appear very pale when viewed under the microscope and consequently give featureless spectra when examined by MSP. A good quality Raman spectrum can be recorded from pigments in such fibers without difficulty because they produce a strong signal.

In paint examination, in addition to the use of high temperature pyrolysis gas chromatography/mass spectrometry (PyGC-MS), low temperature PyGC-MS has been shown to produce excellent results for accurate identification of paint additives. This two-step process has greatly enhanced the discriminating power for paint analysis. In addition, Raman spectroscopy, using a dispersive instrument, has been shown to be useful for the analysis of organic pigments in automotive coatings.

In the face of continuing development of new automotive pigments, it has been necessary to find the best ways of identifying them. The knowledge of which major and minor pigments are present in a paint sample is very helpful in assessing the evidential value of matching samples. Different paint suppliers use different pigment mixtures to obtain the same color shade, so identification of the pigment composition increases the discriminating power. If the dates of introduction and/or composition of certain pigment mixtures are known, this information can greatly assist in identification of the make, model and year of vehicles involved in hit-and-run accidents. The analysis of plate-like pigments in automotive coatings has been discussed in detail by Stoecklein (2001).

In glass analysis, trace element determination using inductively coupled plasma-mass spectrometry (ICP-MS) in conjunction with laser ablation and micro-X-ray fluorescence (μ-XRF) now permits trace element determinations on very small glass particles. This advance is important because it is now possible to distinguish between float glass from different furnaces or plants, resulting in a considerable increase in discriminating power over that achieved as the result of using refractive index measurements alone.

The science involved in the examination of trace evidence has increased beyond measure, as illustrated by the following. In the late 1970s the former Metropolitan Police Forensic Science Laboratory in London (now London Laboratory, Forensic Science Service) took an important step towards standardizing its procedures and produced a *Biology Methods Manual*. The section on fiber analysis methods took 40 pages. In contrast, the Manual of Best Practice for the Forensic Examination of Fibers completed in 2001 by the ENFSI European Fibers Working Group, has nine sections and consists of some 200 pages.

Every 3 years the INTERPOL Forensic Science Symposium takes place in Lyons, France, and selected experts are asked to prepare a review of the developments that have taken place in their subject specialty area since the last conference. The review on paint and glass for the September 2001 conference was prepared by Dr. Wilfried Stoecklein and Dr. Stefan Becker of the German Federal Police Laboratory (Bundeskriminalamt). It contained 169 references to papers published during that time interval, which is almost *60 new papers a year* pertinent to forensic examination of glass and paint (and only the major contributions are listed). The figure for fiber-related publications appearing in the period 1995–1998 was even higher: 218 new references.

RESEARCH AND THE ESTIMATION OF EVIDENTIAL VALUE

Assessment of the evidential value of trace evidence has been made easier during the last 10 years thanks to a considerable amount of research carried out to accumulate background data on frequencies of occurrence. Assisted by the formation of, and cooperation within, international working groups, tasks of considerable logistical magnitude, formerly beyond the resources of single laboratories acting alone, have become much easier to manage. Research providing useful data has often been prompted by problems arising in casework.

A great deal of recent research in the area of fiber examination has concentrated not so much on developing new analytical methods but rather on how to improve interpretation of the case findings.

Target fiber studies designed to establish how often a selected fiber may be expected to appear among a randomly sampled fiber population. At least eight such studies have now been completed, dealing with the main generic varieties of fibers.

Population studies designed to evaluate the content of a randomly sampled fiber population by dividing it into generic type/color combinations. Populations studied so far include those on cinema and car seats, T-shirts and underwear, and outdoor surfaces.

Studies on blocks of color designed to see how frequently certain (dye) spectral patterns occur within a group of samples falling into one generic type/color combination, for example, blue polyester, green acrylic, etc. Most work to date has concentrated on cotton fibers, as comparison of natural fibers is heavily reliant on color comparison.

Live trials designed to assess the fiber transfer/retention potential of casework garments by simulating (as far as possible) the actual conditions of transfer in the case being investigated. This may be problematical, as the exact information may not be available, but in a broad sense the information obtained can be very valuable.

As mentioned previously, research projects often originate as a result of actual problems encountered in casework. The study on outdoor fiber populations above was instigated after a series of cases at the German Federal Police Laboratory involving fiber recovery from various outdoor surfaces. Many of these cases involved terrorist activity. The results can be helpful in assessing the value of findings as they provide "baseline information" about the types and numbers of fibers that might be present on such surfaces, without any specific recent contact having been involved.

Even traces of a very common material, like indigo-dyed cotton fibers which originate from blue denim clothing, may become significant if found in a high concentration on a surface where it would be reasonable not to expect them to be present (on an automobile bumper or under-carriage, for example) except perhaps as isolated examples.

The fibers in a red acrylic scarf played a central role in one of Austria's most famous criminal cases. A target fiber project was undertaken by members of the European Fibers Group in order to assess the significance of the fiber findings in this case. In this project, 435 fiber tapings made from upper outer garments were collected throughout Europe and examined in 38 laboratories in 19 countries for fibers matching those in a red scarf worn by the suspect in a series of murders of 11 prostitutes in Austria, Czechoslovakia, and the US in 1990 and 1991. The case was examined in the Zurich Police Laboratory in Switzerland. In the study, only one indistinguishable fiber was found on a casework garment and one on a personal garment. The project results confirmed the Swiss scientist's assessment of the value of the evidence: the transfer of fibers found in the case was very unlikely due to coincidence.

These fibers were interesting for another reason: they were the first documented example where an additional peak was visible in the infrared spectrum which was clearly attributable to the dye present in the fiber. This finding generated interest in the possibility of identifying fiber dyes using FTIR or Raman

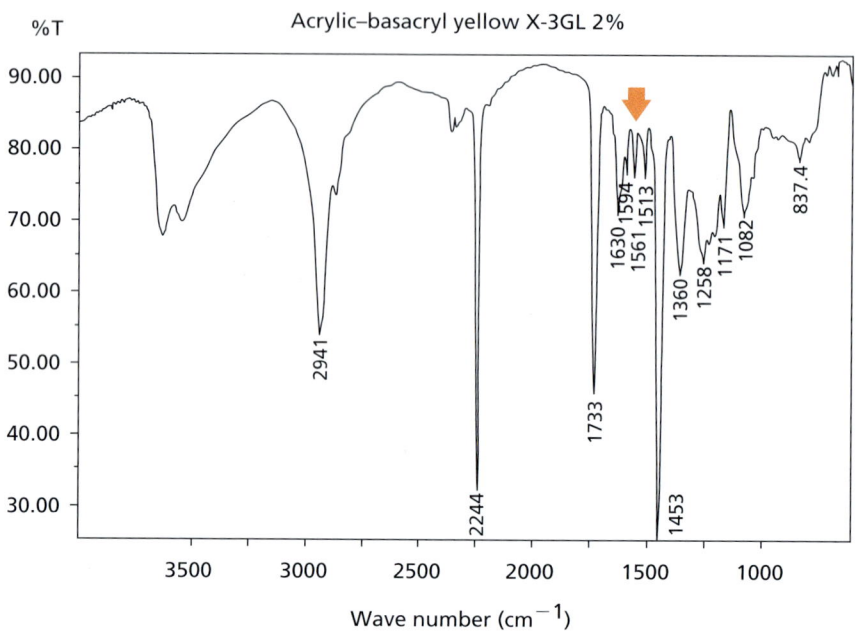

Figure I.8

Spectrum of an acylonitrile/methylacrylate co-polymer fiber showing typical "dye peaks" in the fingerprint region.

spectroscopy and provides a good example of research which can lead to increased sample discrimination being spawned by a casework observation. A typical example of such "dye peaks" appears as in Figure I.8.

The laboratory in Zurich, Switzerland, is one of the leaders in trace evidence examinations in Europe. They have a long tradition to uphold, for the founder and former head of the laboratory was Max Frei-Sulzer (1913–1983), another great name in the annals of trace evidence. Frei-Sulzer, a disciple of Edmund Locard, was fascinated by the use of microscopy from an early age. He was responsible, together with his Swiss colleague Martin (Frei-Sulzer, 1951; Martin, 1966), for initiating the use of sticky transparent tape to recover fibers and other trace evidence from suspect's clothing. Thirty-five years ago, he wrote an article on "Coloured Fibres in Criminal Investigations, with special reference to Natural Fibres" (Frei-Sulzer, 1965) in which he anticipated routine use of microspectrophotometry and chromatography to examine dyed fibers. Even at that time he was concerned with the possibility of laboratory contamination and set up special dust-free rooms for evidence collection.

One of his most interesting cases involved solving a multiple murder by microscopical examination of a crime scene soil sample left from the shoe of the murderer. This case has been described in detail by Palenik (1982) who referred to Frei-Sulzer as "Sherlock Holmes with a microscope". Incidentally, much of the pioneering work on the use of microspectrophotometry for forensic examination of fibers was carried out in the Zurich laboratory by scientists Amsler, Halonbrenner, and Meier. This technique has become absolutely critical for color comparison of both paint and fiber samples, as well as being used for the examination of inks.

THE FUTURE OF TRACE EVIDENCE

Forensic science has undergone important changes in the last decade. In some laboratories, the police now pay money for forensic services. The result has emphasized a cost-driven "need for speed," which has tended toward careful selection of the number of items analyzed and only those likely to yield the most probative information are examined. The cost factor has tended to work against trace evidence, with its lengthy examination times, and in some cases has forced trace evidence analysts to re-assess how they can become more effective (Grieve and Wiggins, 2001).

An article by Petraco in the *Journal of Forensic Sciences* in 1986 (Petraco, 1986) was written, "to encourage further the current resurgence of interest in the use of microscopy for the study of trace evidence in the crime laboratory." Times change. Shortly afterwards, DNA analysis of biological materials began to gain momentum. It found great favor because, apart from being relatively rapid, it offered spectacular precision with its ability to produce odds running into billions when referring to the origin of stains. The term "genetic fingerprinting" captured the imagination of press and public alike especially. Judges and jurors appreciated the "certainty," if not the complexity, of DNA typing, compared to mentally wrestling with the more statistically vague offerings often resulting from trace evidence examinations.

At the turn of the 21st century some forensic science laboratory administrators and forensic scientists are beginning to realize that they may have been a little hasty in forecasting the demise of trace evidence examinations. And, perhaps, they also regret the systematic reduction of personnel, funding, and space to which trace evidence departments were subjected. In more and more cases, DNA analysis plays little to no role whatsoever because shrewd criminals take steps to avoid leaving biological materials behind. DNA analysis has inherent limitations, also. It can only tell whose biological material is present in a certain location if that person's DNA is on record or a known sample was collected. It cannot tell if the right samples were examined or if other exonerating biological material may be present nearby, and it is not possible to determine the time of deposition of the relevant material.

There will always be cases where DNA analysis is unable to assist in the investigation simply because the requisite biological material may not be present:

- Theft through breaking and entering: **paint, glass, soil, fibers**
- Automobile theft: **fibers, hairs**

- Driver determination: hair, fibers, plastic fusions
- Hit-and-run accidents: hairs, fibers, paint, glass
- Construction of terrorist devices/kidnapping cases: **fibers, tape**
- Linking weapons/masks/shared dwellings/vehicles to terrorists and/or armed robbers: **fibers, hairs, soil**
- Providing evidence of drowning, association with aquatic scenes: **diatoms, aquatic ecology**
- Evidence of association with crime scenes: **whole palette of trace materials**
- Providing investigative leads from deceased persons: **fibers, clothing, hairs, soil**
- Identification of the source and movements of illegal materials through trace evidence associated with these items or their packaging: **fibers, hairs, tape, glass, soil**
- Transfer of clothing/household textile fibers and hairs in rape/sexual assault cases where no bodily fluids are involved possibly due to a condom: **fibers, hairs**

An excellent illustration of why it is hard to define topic areas like fiber and textile examination within forensic science is that in one case where it was necessary to determine the degree of product individualization among examples of a childrens' soft toy called "Pikachu"® shown in Figure I.9, which originates from the game Pokemon®.

The recent drive toward quality assurance and accreditation of laboratories and certification of individual examiners has led to the standardization of

Figure I.9

Product variation among "Pikachu"® children's soft toys. Variations can be seen in the shape of the head, the black tips of the ears and the angle of the ears, the arms, the positioning of the eyes and red cheek spots, the angle of the mouth and the length and intensity of the seams and folds.

examination techniques. Much emphasis is being placed on ensuring that examiners are competent and many different ways to measure and register competence have been advanced. In addition to educational requirements set forth in Guidelines there is a trend toward personal certification of examiners. In the US the American Board of Criminalists (ABC) offers certification of trace evidence examiners with a General Knowledge Exam (GKE) and specialty examinations in fibers, hairs, and glass. In the UK, a council for the Registration of Forensic Practitioners (CRFP) has been established with a team of expert assessors in various subject areas who will process examiners' applications for certification. Part of the assessment will include a review of recent casework files completed by those persons seeking certification. The Forensic Science Society issues diplomas in different areas of forensic science and is at present preparing one for fibers examiners.

A Forensic Paint Analysis and Comparison Guideline was published in May 2000. It was produced by the Paint Subgroup of the FBI-sponsored Scientific Working Group for Materials (SWGMAT) to assist personnel who conduct forensic paint analysis in the evaluation, selection and application of tests that may be of value to an investigation. It was intended to be a revision of the original American Society for Testing & Materials (ASTM) E 1610-94 Guide. The content provides an introduction to the forensic examination of paint and coatings. Both the SWGMAT Fibers and Glass Subgroups have published similar guidelines which will also be submitted to ASTM for review, vote, and publication by that Society. As already mentioned, in Europe, the European Network of Forensic Science Institutes – European Fiber Group has just completed a Best Practice Manual for the Forensic Examination of Fibers. This represents the consensus of opinion of scientists from about 50 laboratories in 23 countries.

The compilation of these guidelines has been arduous and time consuming, requiring examiners from many different laboratory systems in different jurisdictions to reach agreement on the procedures and methods that will determine their future. These laboratories vary in staffing, experience, and equipment. It must be remembered that the aim of producing these guidelines is to raise the standards of trace evidence examinations worldwide. Although they can only be advisory, rather than mandatory, in nature, it is crucial that they do not become the lowest common denominator among the parties striving to reach agreement. It does not serve the interests of quality assurance if the level of discrimination achieved during analyses is less than it might be because the best techniques available at present are not being used.

Despite the current tendency to concentrate on improving knowledge about frequency of occurrence, the analytical side of the picture must not be forgotten. All interpretations of the value of the findings in trace evidence cases, whether based on classical statistical methods or the use of Bayesian theory, depend on the fact that the control and recovered samples are found to be "a match."

Therefore, the trend in analytical technology has been to use combinations of methods that provide the best discriminating power between samples. Trace evidence analysts now have a battery of highly sophisticated instrumental techniques at their disposal. Discrimination of paint batches can only be solved if methods like micro-FTIR, pyrolysis gas chromatography, energy dispersive X-ray spectrometry (SEM-EDS) are used together with microspectrophotometry in both transmittance and reflectance. In order to increase the evidential value of glass examinations there has been a recent trend toward increasing use of elemental analysis using ICP-MS, LA-ICP-MS, SEM/EDS, µ-XRF, TXRF, in addition to making refractive index determinations.

Irrespective of the material being examined, the aim of increasing the discriminating power is to reduce the number of groups of samples (classes) which have common characteristics, thus creating more and more distinguishable groups. The ultimate aim would be sample individualization. Some characteristics will occur in the general population much less frequently than others and samples exhibiting these will, therefore, have an increased evidential value. The chance of such an infrequently occurring sample being found in a case and matching the known sample involved, purely by coincidence, then becomes very much reduced.

Research in the field of trace evidence continues to increase as evidenced by the wealth of recent publications. This has resulted in much more background information and data being available to trace evidence analysts to help them with the interpretation of findings. Research and data acquisition have become simplified, not only through the formation and cooperation of international working groups but also because the advances in information technology have made storage and sharing of large volumes of data so much easier.

The use of validated methods and protocols in trace evidence analysis in the USA, Canada, and Europe and the introduction of quality management systems with traceable reference standards for analytical examinations greatly improve the possibilities for international sharing of databases. The new emphasis on quality has led to focusing on specific problems, for example that of sampling at the crime scene – a problem particularly relevant to glass cases.

It is generally agreed that improving the service which trace evidence examination provides is heavily dependent on forensic scientists being better able to interpret casework results and better equipped to communicate their significance to all concerned in the legal process. Considerable progress has already been made in dealing with the issue of sample frequency. One of the biggest needs is to make sure that any data generated is internationally applicable and exchangeable. Communication is no longer a problem. It is essential that data recorded and contributed to central data banks by potential users is compatible, which is where the use of standardized techniques and best practice can bring significant improvements. Perhaps the biggest problem is in providing the time and personnel necessary to keep a database up to date.

However, some other difficulties exist that cannot be completely solved by quality assurance. For example, a key morphological characteristic involved in the assessment of fiber frequency is that of the level of delusterant present in individual fibers. The delusterant is particulate material, usually titanium dioxide, added to the molten polymer, which will affect reflection of light making the fibers appear dull or lustrous. Under the microscope, the delusterant appears as a series of randomly distributed black dots. The problem is that because the particles do not lie in one plane, they cannot be reliably quantified. In side-by-side comparisons with a comparison microscope, this does not represent a problem as small differences in particle size and distribution can be seen without any difficulty. The difficulty is to place each sample into a defined category describing the delusterant level it contains. Only when this has been achieved will it be possible to more accurately research the frequency of occurrence of the individual categories. The creation of data banks designed to provide "intelligence" is more complex and requires greater planning than those designed to assist with, or confirm, sample identification. The foundation for these "reference databases" is generally existent and may come from outside forensic science, but they need to be continually updated.

Some examples of modern forensic trace evidence data bases are:

The Royal Canadian Mounted Police Paint Data Query (PDQ): The PDQ Data base of automotive paints, which make it possible to trace an original finish paint sample to the make/model/year through comparison with the color of undercoat layers, the sequence of the paint layers and the chemistries of each layer. There is an infrared spectral library of all paint layers in the database.

EUCAP – The European Collection of Automotive Paints: This database has been supported by the European Paint Group, a working group of the European Network of Forensic Science Institutes (ENFSI) 42 Laboratories from 21 countries are jointly engaged in supplying analytical data, information and samples from car and paint manufacturers in their respective countries. The coordinating laboratory for the project is the Forensic Science Institute of the German Federal Police (BKA).

The BKA Catalogue Data Base for Fibers: It is based upon records from about 115,000 clothing items taken from mail order catalogues in Germany. It is designed to give information on the frequency of fibers of the various generic types and the frequency of occurrence of fiber type/color combinations both in the general fiber population and in over 90 different categories of clothing.

The Forensic Science Service Fiber Data Base: The Forensic Science Service in England has worked for many years on developing an extensive database on the occurrence of different morphological characters and polymer sub-types in fibers collected from case work garments. The information can also be related to different textile items. Information from this database can be used in conjunction with that from the BKA.

Considerable background information is now available about fiber populations on cinema and car seats, on T-shirts, and on outdoor surfaces. Recent research in Australia has investigated the presence of glass particles on 2008 upper outer garments which showed that the presence of glass particles on randomly selected clothing is likely to be very low. A similar study was carried out on 776 pairs of shoes – only 0.3% had glass particles embedded in both the uppers and the soles, 5.9% had glass in the sole, and 1.9% in the uppers. Detailed experiments have been carried out on the cross transfer of hairs and fibers of subjects who have been wearing a mask. The fund of knowledge relevant to better interpretation of casework results is ever increasing.

Frequency data of this kind are particularly important if the evidence is to be evaluated statistically or probabilistically. In Europe, support is growing for applying the principle of Bayes' theorem to evidence interpretation. The main criticism of a Bayesian approach is that even when used in a broad sense, the degree of subjectivism concerning frequency figures is unacceptable. The Bayesian approach (Taroni *et al.*, 2001) "considers probabilities as measures of belief. As such it enables the combination of objective probabilities, based on data (but requiring subjectivity in the definition of any underlying models) and subjective probabilities, for example transfer and persistence phenomena, for which the certified knowledge and experience of the forensic scientist may assist in the provision of estimates." One of the latest relevant publications (Taroni *et al.*, 2001) which discusses De Finettis subjectivism may finally help to overcome the reluctance of Bayesian critics. It presents a case for scientists to use subjective probabilities in the process of parameter estimation and evidence evaluation. In fact the differences between using a classical statistical approach for evidence evaluation resulting in a scale of probabilities and using the Bayesian approach are not irreconcilable; both strive to achieve a degree of exactness which is unattainable in trace evidence cases. Bayesian methodology (Taroni *et al.*, 2001) does not pretend to get the "true" probabilities; it is an effective method to analyze, criticize, check "coherence" of peoples opinions, and to help them in revising their opinions in a "coherent" way. No more, but not less than that.

Hopefully, after reading *Trace Evidence Analysis: More Cases in Mute Witnesses*, the reader will be left with the impression that the study and use of trace evidence is alive and well. There is certainly no shortage of ideas for further developments in the field which will further increase its applications, efficiency, and usefulness to forensic science and the criminal justice system. Progress, however, requires a certain level of investment. While it is likely that automation will play an increasing role in the analysis, documentation and interpretation of case work somehow sufficient manpower resources must be maintained to ensure that there is a continuing increase in overall standards rather than merely maintaining our specialist knowledge at its existing level and, more important, ensuring that it does not gradually die out through a slow process of neglect and failure to encourage young scientists to follow in the footsteps of experts who have gone before them. Every case is different and trace evidence is an area of forensic science where long term experience can be particularly valuable.

In closing, for those readers interested in pursuing a more detailed study of trace evidence, the authors would like to mention that in addition to the case studies presented in *Mute Witnesses* and *Trace Evidence Analysis: More Cases in Mute Witnesses*, three comprehensive books have recently been published describing forensic examination of fibers (Robertson and Grieve, 1999), paint and glass (Caddy, 2001), and interpretation of glass evidence (Curran *et al.*, 2000) in considerable detail.

REFERENCES

Caddy, B. (2000) Trace evidence – small samples, big problems, in "Problems of Forensic Sciences" *Proceedings of the 2nd EAFS Meeting, Cracow*, pp. 24–37.

Caddy, B., ed. (2001) *Forensic Examination of Glass & Paint*. London: Taylor & Francis.

Curran, J.M., Hicks-Champod, T.N. and Buckleton, J.S., eds. (2000) *Forensic Interpretation of Glass Evidence*. London/Boca Raton, FL: CRC Press.

Frei-Sulzer, M. (1951) "Die Sicherung von Mikrospuren mit Klebeband," *Kriminalistik*, 10/51, 190–194.

Frei-Sulzer, M. (1965) "Coloured fibres in criminal investigations," in *Methods of Forensic Science*, Vol. 4, ed. Curry, A.S. London: Interscience, pp. 144–176.

Grieve, M. and Wiggins, K. (2001) "Fibres under fire." *Journal of Forensic Sciences*, 46(4), 67–75.

Houck, M., ed. (2001) *Mute Witnesses – Trace Evidence Analysis*. New York: Academic Press.

Locard, E. (1930) "The analysis of dust traces," *American Journal of Political Science*, 6, 276–298.

Martin, E. (1966) "New types of adhesive strips and protection of microscopic evidence," *International Criminalistic Police Review*, 200, 200–204.

McCrone, W. (1997) *Judgement day for the Turin Shroud*. Chicago, Illinois: Microscopic Publications Division of McCrone Research Institute.

Palenik, S. (1982) "Microscopic trace evidence – the overlooked clue. Part 2, Max Frei – Sherlock Holmes with a microscope," *The Microscope*, 30, 163–170.

Petraco, N. (1986) "Trace evidence – the invisible witness," *Journal of Forensic Sciences*, 31(1), 321–328.

Robertson, J. and Grieve, M., eds. (1999) *Forensic Examination of Fibres*, 2nd edition. London: Taylor & Francis.

Smith, S. and Fiddes, F.S. (1955) *Forensic Medicine*, 10th edition. London: J & A Churchill.

Stoecklein, W. (2000) "The analysis of new plate-like pigments in automotive coatings," *Paint & Coatings Industry*, 17(9), 48–65.

Taroni, F., Aitken, C.G.G. and Garbolino, P. (2001) "De Finetti's Subjectivism, the assessment of probabilities and the evaluation of evidence: a commentary for forensic scientists," *Science & Justice*, 2001, 41(3), 145–151.

CHAPTER 1

HAIR OF THE DOG: A CASE STUDY

Silvana R. Tridico

Forensic Scientist, Centre of Forensic Sciences, Toronto, Ontario, Canada
(Formerly at the Forensic Science Centre, Adelaide, South Australia)

INTRODUCTION

"… I had grasped the significance of the silence of the dog … someone had been in and … he had not barked. Obviously the midnight visitor was someone whom the dog knew well." Thus deduced Sherlock Holmes in the case of the missing racehorse "Silver Blaze" (Conan Doyle, 1986). In the early hours of 17th August 1994, two partially burnt and battered bodies of a man and a woman were discovered in what remained of their home. A witness later told detectives that he saw a person, accompanied by a dark, medium sized dog running away from the property shortly before the fire started. An inspection of the property and the immediate environs revealed that the couple's Australian Cattle dog, "Ben", was missing. He was found some days later, 6 miles from home. The murdered couple was subsequently identified as Jeremy Torrens and Karen Molloy. The fire and smoke had practically obliterated all chances of finding any evidence that may have linked the murderer to the scene of the crime. The only physical evidence found was a single bloodspot, located on the doorjamb of the back door. During the ensuing investigation, Dennis Molloy, Karen's son from a previous marriage, was interviewed. He gave conflicting accounts not only on his movements on the night in question but also regarding his last contact with the couple and Ben, his mother's dog. He claimed to be 6 miles away from the couple's home on the night in question yet he was captured on a security camera making purchases at a service station 2 miles from the scene. He claimed that he had not had any contact with Ben for several months before his mother's death; yet hundreds of animal hairs were found on and around his bed, on the inside of his car and on one of his sweatshirts. A fortnight after the murders of Karen Molloy and Jeremy Torrens, Dennis Molloy was charged with their murders.

The evidence implicating him with the murders was circumstantial. An eyewitness saw a person with a dog running away from the scene shortly before the fire; another witness would later testify to seeing a person whom she thought was Dennis, in the company of a dog on the morning after the homicides. Neither of these witnesses could positively identify either Dennis Molloy or the decedent's dog. The security tape showed that Dennis was not where he purported to be on the night in question, it did not prove he was a murderer. The motive that led to his mother and Jeremy's deaths was thought to be financial. Investigators believed that Dennis Molloy went to the couple's home, in the early hours of 17th August, to extract money from them. The couple refused his demand and during the ensuing argument Dennis had beaten them to death. He subsequently set fire to their home in order to destroy their bodies and hence conceal their murders. At trial, the prosecution was later to assert that following the murders; Dennis took Ben with him and drove to his beachside home. Dennis subsequently changed his clothing, put Ben in his car and drove some 8 miles to North Adelaide where he abandoned the dog. Given that the evidence implicating Dennis in the murders was circumstantial, investigators hoped that the DNA profile of the single blood spot found at the scene might resolve the question regarding his presence at the crime scene. However, the DNA profiling showed that the blood could have originated from Karen Molloy and likely to have been deposited by the bloodied hand of the murderer leaving the house. The lack of significant DNA evidence meant that the strongest corroborative evidence that would ultimately link Dennis Molloy to the crime scene, would come from over 400 hundred animal hairs and one Australian Cattle dog.

CRIME SCENE

In the early hours of the 17th August 1994 a plaintive female voice was heard to cry out "Don't hurt him! Don't hurt him!" Neighbors subsequently heard raised voices and thudding noises coming from the property. The time was 3.00 a.m.; by 3.30 a.m. the house was ablaze. The local fire department was swift in attending and in extinguishing the fire. A search of the charred remains revealed two bodies on the floor in the main bedroom. The subsequent *post mortems* would show that neither had died as the result of the fire, but as a result of severe head injuries, probably caused by a repeated blows from a blunt instrument such as a baseball bat.

The decedents, Jeremy Torrens and Karen Molloy, were a middle-aged couple who had recently become engaged. Both had been married before and both had children from these marriages.

On the night of the murders, neighbors of the deceased couple told investigators that they were somewhat surprised that Ben had not barked throughout the argument and ensuing commotion. The animal belonged to Karen; following her divorce she took her children and Ben to live with her. Ben was an Australian Cattle dog (Figure 1.1) and was renowned to being very protective of his human family and aggressive towards strangers.

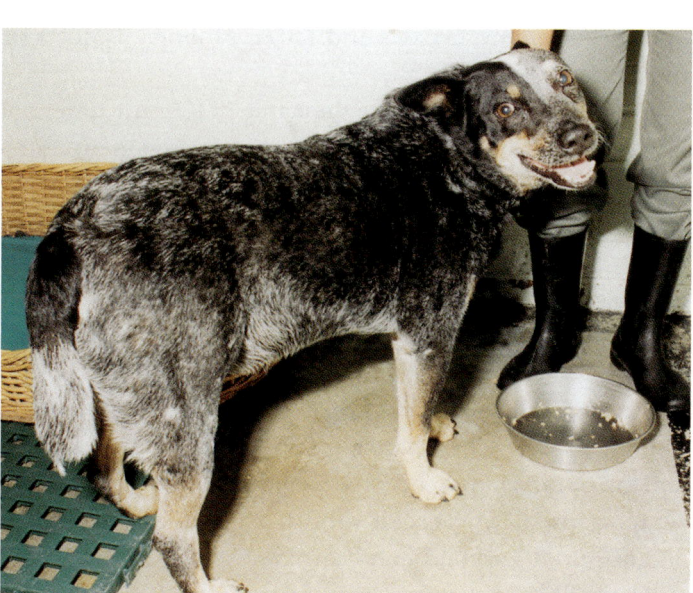

Figure 1.1
The decedents' dog, Ben (Australian Cattle dog cross).

The lack of any evidence to suggest forced entry, the prolonged and vocal argument between the decedents and the intruder, led the police to believe that the murderer was not a burglar but was someone the decedents knew. At the trial, the prosecution reinforced this premise by asserting that the fact Ben remained mute during the murders indicated that he knew the intruder well. One witness told detectives that shortly before the fire he saw a person in the street accompanied by a dog. The hood on the jacket obscured the face thereby making further identification impossible. The darkness precluded further identification of the dog; other than it was medium sized and dark in color.

A search of the crime scene revealed that Ben was missing. He was found a few days later in North Adelaide some 6 miles from Karen and Jeremy's home. A veterinary examination determined that, apart from dehydration, the dog was in good health and that the state of his paws was such that the animal could not have walked this distance. The seat of the fire was subsequently determined to be the main bedroom and forensic analyses determined that the accelerant used to start the fire was petrol.

THE INVESTIGATIVE PROCESS

On the night of the homicides, detectives did not have a suspect. The usual recourse in these circumstances was to begin the investigation with the deceased's family members and friends. In this particular instance the investigators were keen to interview the decedents' family as the evidence suggested that whoever murdered Jeremy Torrens and Karen Molloy was not a stranger but someone they knew.

Neighbors gave the police the name of Karen's 19-year-old son who had been living with his mother until her recent engagement and *de facto* arrangement with Jeremy Torrens. Dennis Molloy by all accounts was a surly and moody individual who was unemployed and relied on his mother for financial support. Family and friends of the deceased couple told detectives that Karen Molloy indulged her son. She bought him expensive designer label clothes and gave him an allowance, which he allegedly spent on alcohol and drugs. As a result of Dennis' reluctance to find employment and acceptance of his mother's constant generosity, Jeremy Torrens' relationship with him was strained. Once he and Karen purchased their home, Jeremy used this as an opportunity to oust the son and encourage him to become independent of his mother. Dennis subsequently rented a unit in a beachside suburb of Adelaide 10 miles from his mother home and acquired a second hand car 2 weeks before her death.

The following day detectives interviewed Dennis as a matter of routine. He claimed to be on good terms with his mother and when questioned on his movements on the night in question he stated that he had frequented the bars of various hotels in Adelaide and returned home around 1 a.m. He stated that he had not seen his mother or her fiancé for several months prior to their deaths.

Following subsequent public appeals for information regarding the homicides, a gas station attendant contacted the police. The gas station was about 2 miles from the crime scene. The attendant provided the police with a security videotape of patrons during the night in question. He thought that the police may be interested in people whom he believed were acting suspiciously. However, an individual, on the tape, who was not acting suspiciously, caught the detectives' attention. At 2.15 a.m. on the 17 August 1994, Dennis Molloy was seen making purchases. The purchases were made using an electronic transaction. Records later showed that he had bought a lighter, a small plastic funnel and 9 l of petrol.

A search warrant for Dennis' residence and car was procured and served. Crime scene personnel found what appeared to be animal hairs on his bedding and inside his car. On the basis of these findings detectives re-interviewed Dennis Molloy and questioned him further on his movements on the night of the homicides and about the relationship he had with his mother and Jeremy Torrens. He was also questioned about his last contact with the family dog and whether the animal had ever been in his car or his unit. The suspect upheld his previous accounts regarding his movements and that he was on good terms with his mother and her fiancé. He further claimed that he had not seen Ben for a number of months and that the animal had never been in his car or at his home. Detectives asked him to explain the security tape evidence and the presence of numerous apparent animal hairs on his bed and in his car. Dennis Molloy refused to answer any further questions or to explain the anomalies in his statement.

Two weeks after the deaths of his mother and her fiancé Dennis Molloy was charged with their murders. Throughout the investigation he maintained his innocence and continued to deny the presence of the family dog, or indeed any other animal, in his car or home. Investigators believed that the forensic examination of the animal hairs recovered from the suspect's environment would be pivotal in placing Dennis Molloy with the dog and hence the crime scene on the night of the homicides.

ANALYSIS

Animal hairs are frequently found during examinations of property and clothing. It is this author's experience, based on many years as a hair examiner, that seldom do the numbers of animal hairs recovered reach double figures. In this investigation, some 400 animal hairs were recovered. These hairs were not distributed randomly over the suspect's belongings; on the contrary they were highly localized. Approximately 300 (80%) of the 400 hairs recovered were found on Dennis Molloy's bedding, one sweatshirt and in his car. The remainder of the hairs consisted of either single hairs or at most, a few found on the remainder of his clothing and in the unit. The distributions of the some 300 hairs recovered from his belongings are given in Figure 1.2. Within his car, over 80% of the hairs were found on or around the front passenger seat. Moreover, the majority

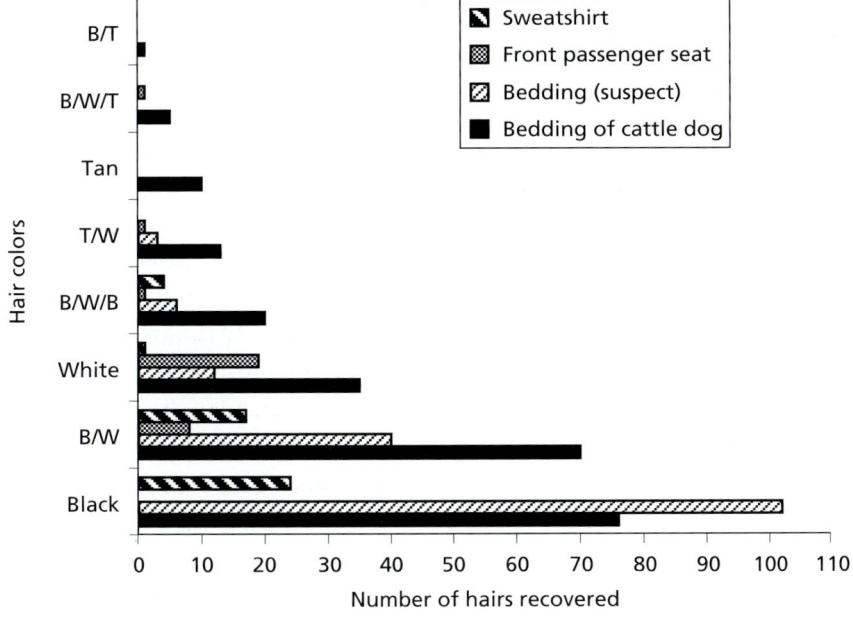

Figure 1.2
Distribution of animal hairs recovered from the bedding of the Australian Cattle dog (Ben) and from items relating to the suspect's environment.

Figure 1.3

(a) Photographs of the front passenger seat of Dennis Molloy's car prior to exerting pressure on the seat. (b) Photographs of the front passenger seat of Dennis Molloy's car after exerting pressure revealing trapped animal hairs in the crevice between the seat and seat back.

(a)

(b)

of these hairs were lodged in the crevice between the back of the seat and the seat itself. These became apparent only on applying pressure to the seat (Figure 1.3a and b). The significance of this finding and the examination of the hairs in the crevice will be dealt with later in the chapter.

The large number of hairs suggested an association of Dennis Molloy's environment with an animal; a finding which challenged his repeated denials that any animal including Ben, had been in contact with his belongings. Dennis Molloy had only acquired his second hand vehicle a few weeks prior to his mother's death. Prior to the examination of the hairs located in his car this author requested that the investigators interview the previous owner to determine if any animal had access to the vehicle. The result of the enquiry revealed that no animals had been in the car.

From a forensic examiner's point of view, the task of examining some 400 animal hairs was somewhat daunting and challenging. The process was likely to be time consuming and would necessitate a systematic and focused approach. The following propositions needed to be considered:

- Hair of a dog? Did the recovered animal hairs originate from a dog?
- Hair of Australian Cattle dog? If the hairs originated from a dog, could that dog have been an Australian Cattle dog like Ben, or could they have come from another breed?
- Were the questioned hairs transferred as a result of the hairs primary transfer, of secondary transfer? Primary transfer results from the animal shedding the hairs directly onto a surface, for example, clothing; secondary transfer would result from the hairs on the clothing being subsequently transferred onto another surface.

The answer to each of these would determine if the examination would continue or cease.

Initially the hairs recovered from the suspect's environment were examined to confirm the observation of the crime scene examiners that the hairs were indeed of animal origin. This confirmation was easily determined, a preliminary examination showed that several hairs exhibited two colors along the shaft, some three; the changes in color being very abrupt and pronounced. This naturally occurring phenomenon known as banding does not occur in human hairs. In addition, examining the roots using a low power microscope which allows the subject to be viewed from 10–50× magnification revealed that the roots of the hairs were large, and elongated, unlike human hairs whose roots are indistinct and round in comparison. These characteristics identified the hairs as non-human in origin. Having established that the recovered hairs were of animal origin, further more detailed analyses were required in order to establish the likely species of origin.

The hairs on an animal consist of a number of hair types that may be characterized on their gross morphological appearance (Appleyard, 1978; Brunner and Coman, 1974). Close examination of animal pelts will reveal that some sparsely distributed hairs are distinctly longer than the remainder of hairs comprising the pelt. These longer hairs are called "overhairs", for species identification purposes these hairs are not particularly useful as many of the characteristics required to determine species may not present or not visible if the hairs are heavily pigmented. The "guard hairs" are coarser and larger than the remainder of the hairs forming the pelt. The largest of the guard hairs, the "primary guard hairs", are of paramount importance in the identification of species as they exhibit the most diagnostic features. The "secondary" guard hairs bear characteristics which are a hybrid of the coarse guard hairs and the finer underhairs, that is they are almost as coarse and the same length as the primary guard hairs and wavy

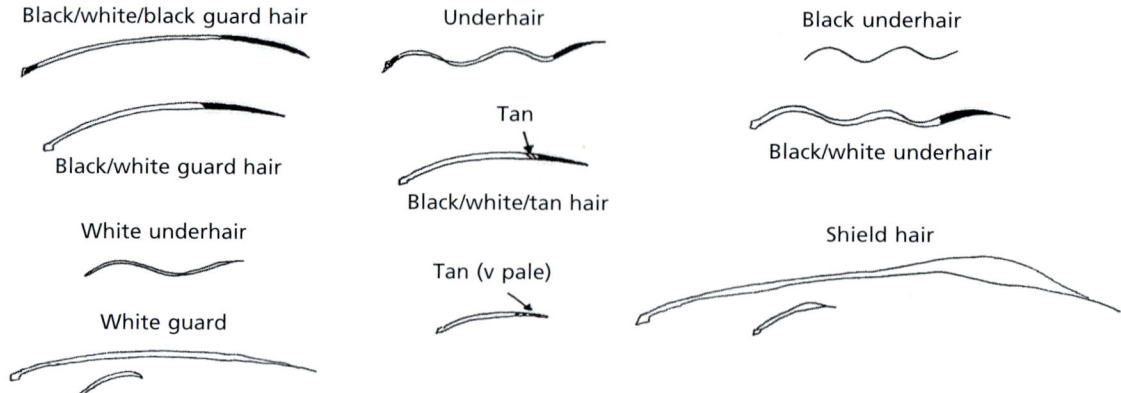

Figure 1.4
The variety of hair types found in the reference sample from Ben (Australian Cattle dog cross) and animal hairs recovered from Dennis Molloy's environment.

like the underhairs. These secondary hairs also bear diagnostic features and are just as important as the primary guard hairs for identification purposes. The "underhairs" are shorter and much finer than the overhairs and guard hairs. These hairs are usually wavy and usually bear insufficient features upon which a reliable identification can be made.

The hairs recovered from Dennis Molloy's environment were received in numerous packages. Each package contained hairs of varying lengths, texture, profiles and colors; clearly, before further work could be performed the recovered animal hairs required sorting into manageable proportions. This sorting process involved placing each of the 400 hairs from their locations into categories comprising guard hairs, secondary hairs and underhairs. Within each of these categories the hairs were further sorted on the basis of color, that is whether the hair was all of one color or banded. In addition, as the hairs were being sorted a tally was kept of all the color types present and if banded, the proportion of each color on the hair shaft was measured and recorded. The hairs were placed into zip lock plastic bags and labeled with the hair type and location found. The range of hairs found and the categories in which they were sorted are shown in Figure 1.4. The sorted hairs, within each of the categories, could not be distinguished either by the naked eye, or by low power microscopy.

Clearly it was not possible to subject all of these hairs to detailed examination. Approximately half the total number of hairs recovered, some 200, within each category was selected for the further more detailed and painstaking examinations.

HAIR OF A DOG?

The majority of the 200 hairs selected for detailed examinations were of the guard hair type, primary and secondary, few comprised some of the finer underhairs. To establish whether the recovered animal hairs could have originated from a dog, a number of features were examined. The roots, scale pattern, medullary

Figure 1.5
Spade shaped root exhibited by many of the recovered animal hairs recovered from Dennis Molloy's environment.

index were examined. Each of these characteristics are diagnostic for a particular species and the reader is directed to the standard works for further detailed descriptions (Appleyard, 1978; Brunner and Coman, 1974; Wildman, 1954).

The majority of hairs examined exhibited spade shaped roots (see Figure 1.5). The appearance of this structure resembling a closed umbrella is typical of dog hairs (Hicks, 1977, pp. 38–39).

The outside of a hair shaft is comprised of overlapping structures, situated like tiles on a roof. In animal hairs these overlapping scales can exhibit a variety of patterns characteristic for a particular species (Appleyard, 1978; Brunner and Coman, 1974; Wildman, 1954). The patterns may or may not be uniform along the hair shaft. In order to view these patterns a scale cast is made. This is performed by coating a microscope slide or cover slip with a thin layer of clear nail polish, the hair is embedded in the nail polish and removed once the polish has hardened. Removal of the embedded hair reveals the scale pattern exhibited by that particular hair. The recovered guard hairs exhibited scale cast patterns typical of dog, that is an irregular mosaic pattern at the root end, alternating diamond like patterns (pectinate)/irregular mosaic mid shaft, culminating in fine wavy patterns at the tip portion of the hair (nomenclature according to Brunner and Coman, 1974). An example of the scale patterns seen along the shaft of dog guard hair is shown in Figure 1.6. In contrast to the guard hairs, dog underhairs exhibit a uniform diamond pectinate pattern from the root to the tip part of the hair, whence the pattern is similar to that seen in the coarser hairs. The underhairs recovered from Dennis Molloy's environment exhibited this diamond pectinate configuration (see Figure 1.7).

During the course of the examination, some guard hairs did not exhibit scale patterns typical of dog. Instead of exhibiting the scale patterns detailed above, these hairs showed a uniform crazy paving-like pattern. In all other aspects these hairs showed features characteristic of dog hairs. By examining other known dog hairs, including those from Ben, this crazy paving pattern was seen in the hairs taken from the dog's abdomen.

An additional characteristic, which may be used to differentiate between cat and dog, is the determination of the medullary index cited in the work of Peabody

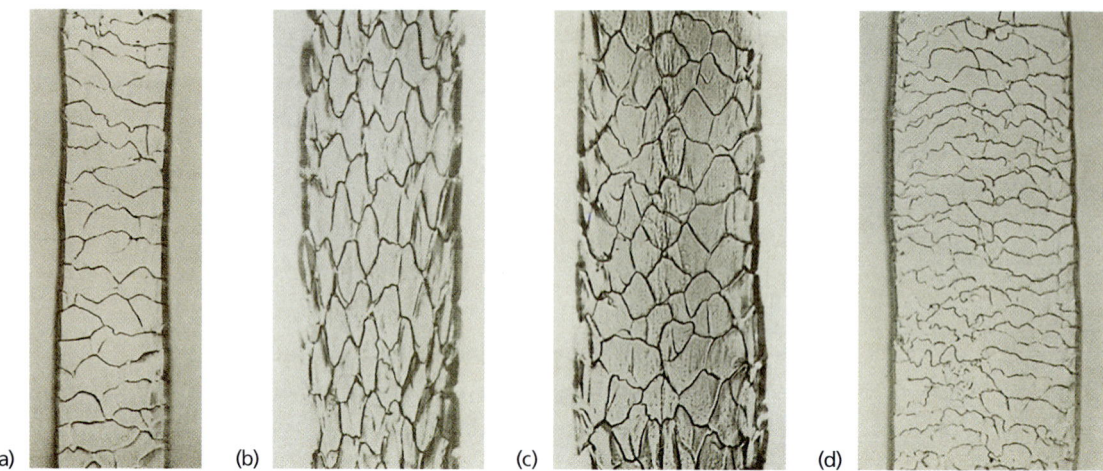

Figure 1.6
Scale cast patterns observed on many of the guard hairs recovered from Dennis Molloy's environment. (a) Root region, (b) along the length of the fibre, (c) along the length of the fibre, alternating with (b) and (d) tip region.

Figure 1.7
Scale cast pattern observed on many of the underhairs recovered from Dennis Molloy's environment.

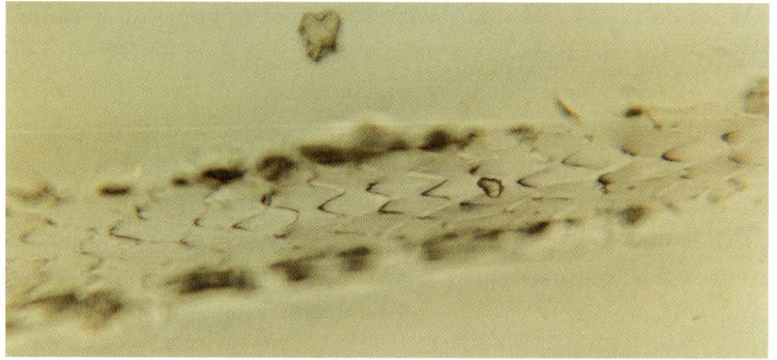

et al. (1983). The medulla is a characteristic present in almost all mammalian hairs, including humans. This structure is an airspace found in the center of the hair shaft and orientated like the graphite lead in a pencil. The medulla usually appears black when viewed under the microscope. In cats the medulla is usually significantly wider than that seen in the dog. Peabody *et al.* calculated ratio of the medulla width to the shaft width (known as the medullary index) for a number of hairs taken from these two animal species. A statistical analysis of the results produced a theoretical line in which the medullary indices of the cat hairs were significantly higher, and were above the line, than those from the dog hairs that were below the line. The medullary indices for the majority of the 200 animal hairs were below this theoretical line (see Figure 1.8). This finding further strengthened the premise that these hairs were of dog in origin.

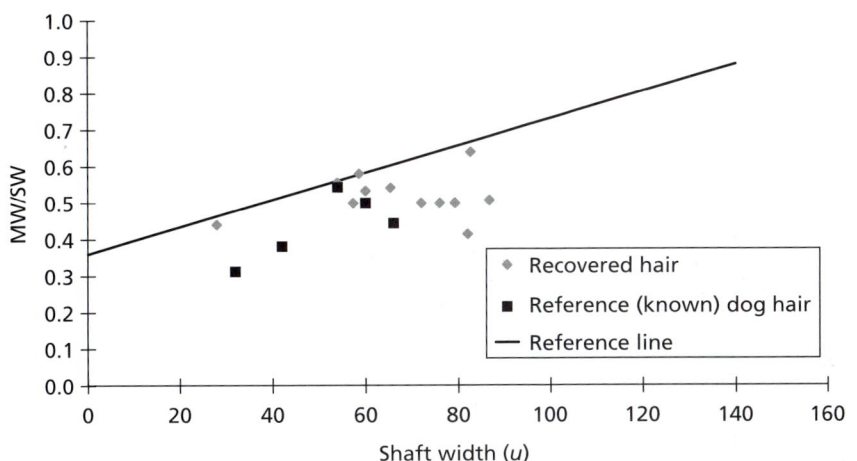

Figure 1.8
Medullary indices of a selection of recovered animal hairs and reference (known) dog hairs.

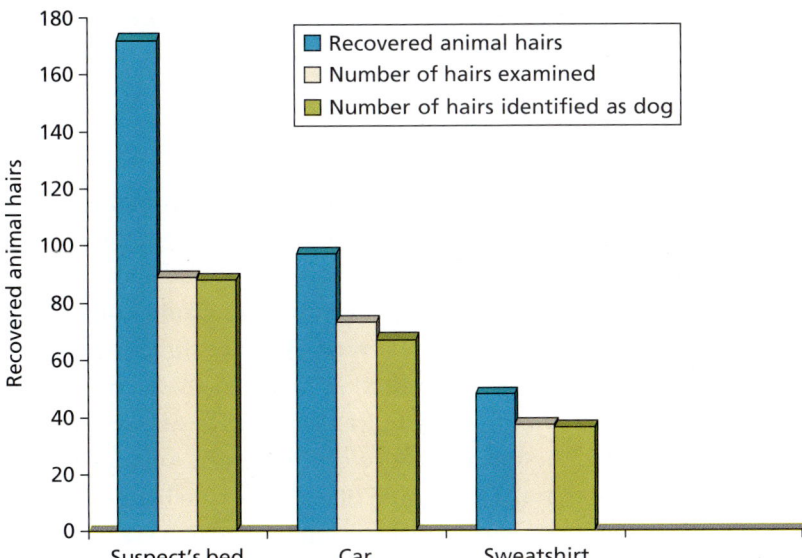

Figure 1.9
Results of the examination of the animal hairs recovered from the suspect's environment.

The results of the examinations established that the majority of hairs recovered from the suspect's environment were likely to have originated from a dog (see Figure 1.9). Similar information regarding the hairs found in the suspect's car is given in Figure 1.10.

HAIR OF AUSTRALIAN CATTLE DOG?

Having identified the 200 hairs as most likely, dog in origin the next key issue to address was whether the source of these hairs could be an Australian Cattle dog, like Ben or from another breed(s).

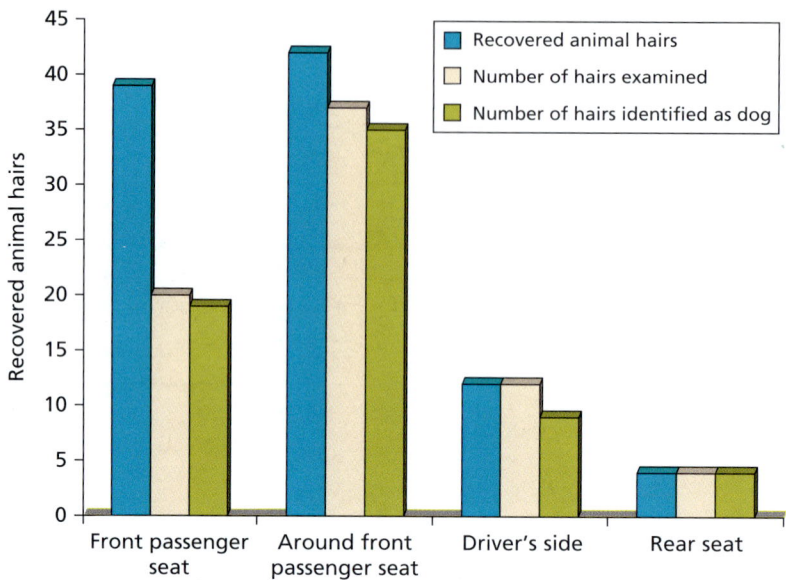

Figure 1.10
Results of the examination of the animal hairs recovered from the suspect's car.

In order to explore whether breeds, other than Australian Cattle dogs, possessed the range of colors present in the recovered animal hairs, the author visited a number of animal shelters, dog boarding kennels, the "South Australian Dog Breeder's Association" and attended several dog obedience classes. No details of the case or of the involvement of an Australian Cattle dog were discussed, a selection of hairs were shown at the various establishments with the question being asked of them was what breed(s) of dog is likely to encompass these hair types. On each occasion the answer given, without hesitation, was an Australian Cattle dog. Indeed, the representative of the "South Australian Dog Breeder's Association" stated that the bi-colored black and white banded hairs and the tri-colored black, white and tan banded hairs in particular were typical of Australian Cattle dogs. Other dog breeds whose hairs bore similar colorings to those found in Australian Cattle dogs were German Shepherd dogs, beagles, collie dogs and King Charles cavalier spaniels. Hairs were taken from a number of these dogs, but none of the hairs exhibited the gamut of colors or the lengths found comprising the unknown dog hairs or those taken from Ben. None of these dogs bore the black, white and tan, tri-colored hairs. A total of 200 dogs of various breeds were examined. Known samples were taken from any dog, irrespective of breed, whose pelt comprised hairs similar to the recovered hairs and those of Ben. The only breeds that possessed the complete array of hairs found in the recovered animal hair samples and those taken from Ben were Australian Cattle dogs.

To determine if the recovered hairs were comparable to the hairs comprising Ben's pelt, a number of hairs were taken from him. In an attempt to capture the range of colors and hair types comprising Ben's pelt, the hairs were plucked from the following body areas the head, chest, upper and lower front and rear legs, abdomen, upper and lower sides of the tail and around the base of the tail. As seen previously in Figure 1.1, Ben was predominantly a black and white dog, his legs and face bore tan colored hairs and the hairs comprising the underside and base of his tail were predominantly white.

These hairs were separated into categories in the same manner as the recovered hairs. The various hair types comprising each category were shown previously in Figure 1.4. Root shape, scale cast patterns and medullary indices were determined using the same processes applied to the examination of the recovered hairs. The roots were spade shaped. The scale cast patterns produced from the guard hairs were an irregular mosaic pattern at the root end; alternating diamond (pectinate)/irregular mosaic mid shaft. The underhairs exhibited a uniform diamond pectinate pattern from the root culminating in fine wavy patterns at the tip portion of the hair. The medullary indices were below the theoretical line a feature typical of dog hairs. These medullary indices were comparable to those obtained with the questioned dog hairs. These results reinforced the conclusion that the recovered animal hairs originated from a dog and that on the basis of the features and characteristics examined were comparable with the hairs taken from Ben. The latter premise was reinforced by the results obtained from measuring the colors comprising the banded hairs. The proportions of each color on the recovered and known banded hairs were almost identical. Selections of the recovered hairs and those from Ben that were tan in coloration were examined using a comparison microscope. This microscope is one in which two microscopes are joined by an optical bridge that enables the simultaneous examination of hairs, up to 400× magnification. The tan colored hairs from both Ben and the recovered hairs were very similar when viewed with the comparison microscope (see Figure 1.11a and b).

PRIMARY TRANSFER VERSUS SECONDARY TRANSFER

The manner in which the animal hairs were transferred onto the bedding, sweatshirt and onto the front passenger seat of the car became an issue at trial. At the time of the investigation no studies had been conducted on the transfer of animal hairs. The conclusion expressed as the most likely mode of transfer of the animal hairs found on Dennis Molloy's bedding, sweatshirt and on the front passenger seat of his car was primary transfer. This opinion was expressed as a result of this author's many years' experience as a hair and fiber examiner. This experience was used in conjunction with the findings conducted on the transfer of fibers, including woolen ones (Lowrie and Jackson, 1994) and on the work conducted on the secondary transfer of human scalp hairs (Gaudette

Figure 1.11
Comparison of the tan portions of the recovered animal hairs (L) and known hairs from Ben (R). (a) A comparison between underhairs. (b) A comparison between guard hairs.

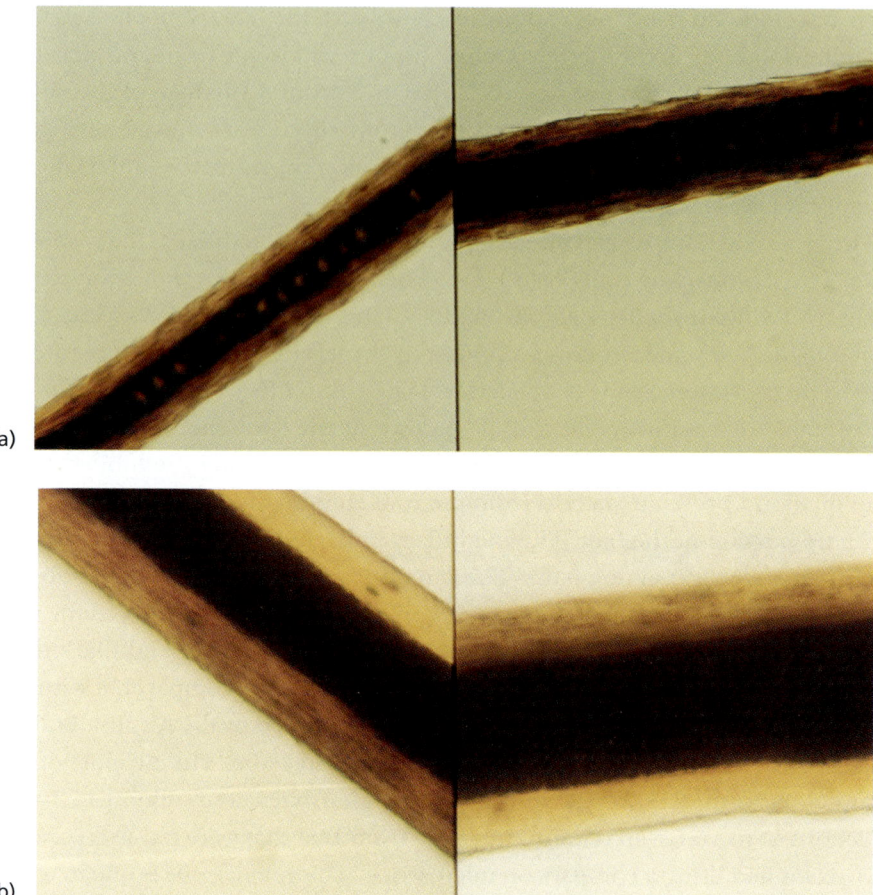

and Tessarolo, 1987). The large number of animal hairs found at the above locations, approximately 170 on the bedding, 46 on the sweatshirt and approximately 85 on and around the front passenger seat suggested primary transfer rather than secondary. The hairs recovered from the bedding comprised overhairs, primary and secondary guard hairs and underhairs. These are the types of hairs one would find on the pelt of an animal. It is this author's experience that secondary transfer is unlikely to result in the simultaneous transfer of all these hair types. This finding resulted in the following examination. In order to gain some idea of the types and numbers of hairs Ben was naturally shedding, his bedding was analyzed. Ben slept in the laundry and although the property had been severely damaged by the fire and smoke, this area was relatively unscathed.

Three, separate and random tape-lifts were taken from the bedding. The tape-lift collecting procedure that is commonly used to collect extraneous hairs and fibers and comprises a strip of adhesive tape being placed on a surface to remove any adhering extraneous material may be removed. The hairs from each tape-lift taken form the dog's bedding were placed into the same categories as those previously performed for the recovered and known animal hairs. The hairs within each category were counted and the distribution compared with the hairs recovered from the suspect's car, bedding and sweatshirt. The results showed a very close correlation between the distribution of hairs recovered from the Cattle dog's (Ben) bedding and those recovered from the suspect's environment (see Figure 1.2). This finding further supported the premise that the hairs on the suspect's bedding resulted from primary rather than secondary transfer, since it would be unlikely that this distribution of hairs would be secondarily transferred. The most predominant color of hairs recovered from the crevice in the passenger seat was white. The significance of this result is discussed in the following section.

INTERPRETATION

The case implicating Dennis Molloy to the murders of his mother, Karen Molloy and her fiancé, Jeremy Torrens was circumstantial. Viewed individually, each piece of circumstantial evidence revealed an incomplete picture; however, viewed *in toto* the picture became increasingly closer to completion. The forensic examination of the animal hairs recovered from Dennis Molloy's environment played a significant role in completing this picture. In establishing this significance it is worth presenting all of the circumstantial evidence collated during the investigation.

In the early hours of 17 August 1994 a heated argument between Karen Molloy, Jeremy Torrens and an intruder was heard. The back and forth nature of the argument suggested that the couple knew the intruder. Ben, Karen Molloy's Australian Cattle dog, renowned for his antagonism toward strangers and fiercely protective of his human family, remained silent this commotion. At trial prosecution suggested that Ben's silence implied that the visitor was someone who the dog knew well. Shortly thereafter, the property was set alight, a witness saw a person, accompanied by a dark, medium sized dog running away from the scene. A search of the crime scene revealed that Ben was missing. A few days later he was found in North Adelaide, several miles from home. The veterinarian opined that Ben could not have walked this distance. This finding suggested that following the murders he was taken, by some form of transportation to the place where he was ultimately found. Several hours after the murders another witness would testify that she saw a person whom she assumed was Dennis Molloy, in the company of a dark, medium sized dog standing outside the back door to Dennis' residence. Neither of these witnesses could positively identify Dennis or Ben as the person and the dog they saw.

The forensic evidence collected from the fire-ravaged property was scant. A single bloodspot on the rear door was analyzed, the DNA profile corresponded with the profile obtained from the blood of Karen Molloy. Fire investigators determined that the seat of the fire was in the main bedroom, where the bodies of the murdered couple were found. Forensic analysis of the accelerant used to start the fire was determined to be petrol.

On the night of the murders Dennis Molloy claimed that he was 6 miles from the scene of the crime. However, the service station security camera proved otherwise; it placed him 2 miles from the scene purchasing, a small plastic funnel, a lighter and 9 l of petrol, some 45 min before the murdered couple's property was set alight. This evidence suggested that either Dennis was mistaken on his whereabouts on the night in question or he was not telling the truth, it did not place him at the crime scene.

The subsequent finding of hundreds of animal hairs in Dennis Molloy's environment contradicted his denials of any recent contact with Ben or that any animal had ever been on his property. These animal hairs were irrefutable evidence that challenged Dennis' credibility. The distribution of these hairs was not random, but highly localized. The vast majority of these hairs were found on his bedding and in the front passenger seat of his car.

The forensic examination of such a large number of animal hairs recovered from the suspect's environment required a systematic and focused approach. The preliminary examination confirmed that the hairs were animal in origin; the presence of banded hairs and large spade shaped roots precluded the hairs as being human in origin. Following the various sorting processes detailed examinations were performed on 200 animal hairs, approximately half the sum total recovered. The results revealed that the majority of these 200 hairs were identified as dog in origin. The classification was based upon a number of features; the spade shaped, elongated roots and the scale cast patterns were typical examples of the characteristics features seen in dog hairs (Appleyard, 1978; Brunner and Coman, 1974). These results when combined with their medullary indices reinforced the premise that the species of origin of these animal hairs was dog (Peabody *et al.*, 1983).

The comparison of the tan colored hairs between the recovered hairs and similar hairs from Ben were comparable, as previously shown in Figure 1.11a and b. The results of the tests conducted on Ben's bedding, showed that the proportions of hairs shed was very similar to those found on the suspect's bedding, on the front passenger seat of his car and his sweatshirt. The dog hairs found in the crevice of the front passenger seat and seat back were only discovered when pressure was applied to the seat (see Figure 1.3a and b). The most natural scenario on the manner in which these hairs were deposited being that a dog sat on the seat and once the animal the got off the shed hairs became trapped in the crevice. The hairs found at this location were predominantly white, the hairs from the other items relating to the suspect's environment were predominantly black (see Figure 1.2). The color of hairs around Ben's backside area including the base of his tail was predominantly white. On the basis of the results obtained from the microscopic comparison of the tan hairs, the distribution of hairs on Ben's bedding and on the hairs found on the front passenger seat the author concluded that he could not be excluded as a possible source of the recovered dog hairs.

The examination of hairs taken from various breeds of dog showed that the only breed that exhibited the entire spectrum of hair colors found in the suspect's environment was the Australian Cattle dog. Indeed, in the opinion of the representative at the "South Australian Dog Breeders Association" the black and white and tan-banded hairs were very typical of hairs found on Australian Cattle dogs. These findings suggested that an Australian Cattle dog could have shed the hairs found on Dennis Molloy's property.

The mode in which the dog shed the hairs onto Dennis Molloy's property, in particular onto his bedding, was a contentious issue at the trial. In the author's opinion the most likely mode of transfer resulted from a dog shedding its hairs by coming into direct contact with these surfaces (primary transfer) and not as a result of the dog hairs being shed from an intermediary object such as clothing (secondary transfer). This interpretation was made on the basis of the large number of hairs found on the suspect's property. At the time of the investigation no published data existed regarding the transfer and persistence of animal hairs. In this author's experience the finding of hundreds of animal hairs was not commonplace and was seldom seen. This premise was supported by Lowrie and Jackson "… most forensic scientists who deal with fibers would agree that a large number of transferred fibers suggests recent direct contact" (Lowrie and Jackson, 1994, p. 74). This author believed that woolen fibers would be the closest analogous fiber to hairs as they were animal in origin and possessed scales on the outside of the hair shafts. Thus, it was reasonable to assume that the results of the studies on the transfer and persistence of woolen fibers could be extrapolated to give an indication of the manner in which shed dog hairs would behave under similar circumstances. Freckleton and Selby (pp. 8-3921 and 8-3925) supported this premise,

> "there are elements common to fibers and hairs (after all, hairs are just natural fibers and the factors which affect their transfer and persistence appears similar) …. Most of the work on transfer and persistence relates to studies with textile fibers. Hairs could be expected to behave like woollen textile fibers because both have scaled outer layers. Hence, many of the conclusions reached for fibers may also be applicable to hairs."

Lowrie and Jackson (1994, pp. 74 and 80) further stipulated that

> "experience suggests that the numbers of fibers transferred by secondary routes must generally be less than those transferred through primary routes.... In the ideal conditions of this study less than 10% of donor fibers present on the primary recipients were transferred to the secondary recipient in the majority of experiments"

These authors cited that similar studies conducted by Grieve *et al.* (1989) supported this finding by concluding "that secondary contact via clothing is likely to result in considerably fewer fibers than a primary transfer" (Lowrie and Jackson, 1994, p. 80). Gaudette and Tessarolo concluded that one of the factors that appears to make secondary transfer less likely than primary transfer is that if a large number of hairs is transferred, it is unlikely that they would all be secondarily transferred because the mechanisms of transfer are more complicated (Gaudette and Tessarolo, 1987).

Furthermore, the results of the distribution hairs Ben shed onto his bedding and those recovered from Dennis Molloy's bedding were almost identical. This distribution would unlikely to have occurred as a result of secondary transfer. The interpretation of the manner on which the hairs were deposited on the suspect's property did not conclude at the completion of the forensic examination but continued into the witness box. It is worth discussing aspects of the trial that focussed on the forensic examination of the animal hairs.

THE TRIAL

The trial was held in Adelaide, South Australia, on December 1995. In scrutinizing the forensic evidence concerning the animal hairs, defense sought the assistance of two hair experts. Prior to trial the defense maintained that Dennis Molloy did not have recent contact with Ben. Following this author's presentation regarding the forensic examination of the animal hairs, the defense conceded that Ben could be the source of the hairs found on Dennis Molloy's bedding, car and sweatshirt. They stated that Karen Molloy, with Ben, visited Dennis on the Friday prior her death. During this visit the dog may have jumped into his vehicle and onto the front passenger seat. Defense also put forward the proposition that Ben's hairs were deposited on the sweatshirt as a result of Dennis' contact with the dog on a prior visit to his mother's home.

However, the defense contested the premise that the most likely mode of transfer of Ben's hairs onto Dennis' bedding resulted from primary transfer rather than secondary. Defense presented the scenario that the hairs were secondarily transferred from a baby blanket, a blanket Dennis asked his mother to bring

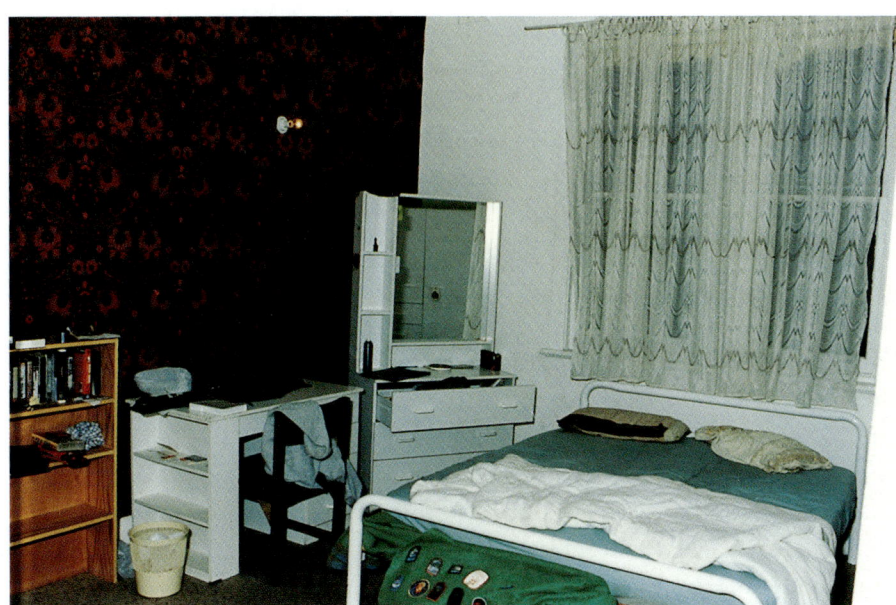

Figure 1.12
Bedroom of Dennis Molloy showing the bed and green baby blanket on the ottoman at the foot of the bed.

with her last visit prior her death. Defense proposed that during the 45 min trip, from her house to Dennis', Ben had sat on the blanket. The blanket was subsequently placed on his bed. Thus, Ben's hairs transferred onto his bedding came from the baby blanket and not from the dog himself. Crime scene examiners did not collect the baby blanket during the examination of Dennis' bedding, the small blanket was on an ottoman at the foot of the bed (see Figure 1.12) and no animal hairs were apparent. The green acrylic blanket was examined during the trial and it bore fewer than 10 white primary guard hairs. On the basis of the number and the types of hairs on the blanket, the defense's proposition that the hairs on Dennis Molloy's bedding were secondarily transferred from the baby blanket was rejected by the author. A whole range of colors and types of dog hairs were found on Dennis Molloy's bedding. The hairs found on the baby blanket comprised exclusively of white guard hairs. If the hairs on Dennis' bedding were shed from the baby blanket then it would be reasonable to assume that not all of the colored hairs would have been transferred and some should have remained on the baby blanket. In addition, the types of hairs and the proportions of hairs that were found on Dennis' bedding were very similar to the hairs shed naturally by Ben (Figure 1.2). These findings supported primary, rather than secondary, transfer.

Studies performed on the secondary transfer of fiber and human scalp hairs (Gaudette and Tessarolo, 1987; Lowrie and Jackson, 1994) indicated that a large number of recovered fibers indicated primary transfer and not secondary transfer. In addition, Lowrie and Jackson, concluded "In the ideal conditions of this study less than 10% of donor fibers present on primary recipients were transferred to the secondary recipient (Lowrie and Jackson, 1994, p. 80). Using this figure, if the baby blanket was the source of some 170 dog hairs found on Dennis' bedding, then the blanket should have had approximately 1500 dog hairs remaining on it and not the extremely low number seen. One of the hair examiners, acting on behalf of the defense, challenged the premise that the animal hairs were likely to have been deposited on the suspect's bedding as a result of primary transfer. The basis of the challenge was that no scientific studies or experiments had been conducted to assess the primary and secondary transfer of animal hairs. Therefore, this author's evidence was anecdotal and not based on scientific fact. During cross-examination by the prosecution, this witness agreed that a *prima facie* case had been established for that the hairs on Dennis Molloy's bedding were more likely to have been deposited as a result of primary transfer, rather than secondary transfer.

Dennis Molloy was found guilty of the double homicides and sentenced, on appeal, to life imprisonment with 25 years non-parole period.

SUMMARY

The investigation into the murders of Karen Molloy and Jeremy Torrens emphasized the value of physical evidence in forensic science. In the absence of significant DNA results, fingerprints or eyewitnesses the examination of these animal hairs was the cornerstone of the circumstantial case implicating the son of Karen Molloy in the homicides. However, this statement is akin to beginning reading a book with the final chapter. A number of events occurred which made the examination of the animal hairs recovered from the suspect's environment viable.

The investigators and the crime scene personnel recognized the potential value of the animal hairs as evidence and were diligent in collecting and storing the hairs in an appropriate manner. The training and education of crime scene examiners in South Australia, regarding the collection of forensic evidence, was based on the G.I.F.T. principle – Get It First Time, that is, all of the available evidence should be collected from the scene or from items. As the investigation progresses it may become clear that the examination of the physical evidence is not warranted, in which case, nothing is lost. Scenes and items should be treated as though the case will depend on physical evidence. The collection of hairs, from whatever surface they are adhering to, should be performed in an appropriate manner, one that preserves their integrity and ensures that the evidence is not contaminated. In this investigation had the crime scene personnel not used the G.I.F.T. principle or had incorrectly dealt with the collection of the hairs found at the suspect's unit and car, then, the forensic examination of the animal hairs would not have been an option. The Forensic Science Centre in South Australia, following the implementation of DNA profiling, continued to maintain a strong commitment to the examination of physical evidence, such as hairs. Accordingly, the hair expertise existed in order to advise and assist the investigators with enquiries regarding the examination of the animal hairs including the collection of known samples from Ben. The advice given to the crime scene personnel to pluck hairs from several areas of his body and not be limited to the dorsal hairs only proved invaluable. The scale cast patterns produced from some of the recovered hairs did not correspond with the patterns depicted as typical for dog hairs (Appleyard, 1978; Brunner and Coman, 1974). However, the scale cast patterns made exhibited by the hairs collected from Ben's abdomen were comparable to the patterns exhibited by the recovered hairs. This finding emphasizes the need for the hair examiner to be vigilant when dealing with the examination of animal hairs and, while appreciating the enormous value and role the literature affords regarding the identification of animal hairs, to be aware that they are not definitive or exhaustive works. According to Brunner and Coman "It is important to realize that the photographs in this section represent only some of the multitude of hair structures observed in the hair of any one species. Generally speaking, only the most diagnostic features have been covered." (Brunner and Coman, 1974, p. 19).

At the time of the investigation DNA profiling of animal hairs was not an option, the technology did not yet exist. However, over the last few years studies have been conducted to evaluate the DNA profiling of animal hairs for forensic purposes (Fridez *et al.*, 1999; Menotti-Raymond *et al.*, 1997; Savolainen *et al.*, 1997, 1999); the results are promising. Indeed, in 1996 DNA typing results of cat hairs played a role in convicting a defendant of second-degree murder (Menotti-Raymond *et al.*, 1997). The success of the DNA profiling is dependent upon the correct species primer being used for the amplification process. The correct species classification of an animal hair, prior to DNA analysis, will be reliant upon the microscopic examination of the morphological features exhibited by the hair. Although, the DNA profiling studies conducted on animal hairs are yielding promising results, this technique is not one routinely used in the forensic arena. Until such times that the forensic DNA profiling of animal hairs is routine, investigators' only recourse will be to rely on the "traditional" forensic analysis of animal hairs. Hair examination can be an extremely time consuming task, indeed the examination of the animal hairs in the double homicide investigation took some 9 months to complete. In the face resource constraints, many laboratories have reduced or abandoned their commitment to these activities. Once the skills and experience are lost it is very difficult to recover lost ground. This case clearly illustrates and emphasizes the value of maintaining these traditional forensic skills.

THE "ROOT" OF THE PROBLEM

The most frequently encountered animal hairs during forensic examinations are those originating from cats and dogs. This is probably due to the popularity of these animals as domestic pets and the relative ease with which these animals shed their hairs onto clothing and upholstery. The differentiation between hairs originating from cats or dogs may be easy, either the hair examiner is sufficiently experienced or the hairs are particularly distinctive. As Moore noted, however, "the identification of animal hairs is generally recognized to be one of the more difficult types of examination" (Moore, 1988). The reason being that variation may exist not only within species but also between the hairs from the same animal. Accordingly, before a classification can be made a number of features need to be examined. The standard texts of Brunner and Coman, and Appleyard, present photographic documentation and basic keys to assist the examiner in the differentiation between cat and dog hairs (Appleyard, 1978; Brunner and Coman, 1974). These works mainly depict the various scale patterns exhibited by hairs from a particular species. Moore produced a key for the identification of a number of animal hairs in which she observed that one of the distinguishing features between cat and dog hairs is that the medulla is wider in cat and narrower in dog (Moore, 1988). Peabody *et al.* took this observation one step further by measuring the width of the medulla in relation to the shaft width to produce a medullary index (Peabody *et al.*, 1982). The aim of the study was to use a statistical method in an attempt to provide some measure of the reliability of the classification when used in conjunction with the morphological features exhibited by the putative cat or dog hair. The results were presented graphically as a straight line separating the medullary indices derived from cat hairs, from those derived from dog hairs. In general, if the medullary index of a putative cat or dog hair was above the line, then the hair was more likely to have originated from a cat than a dog. Hicks (1977) stated that one of the general features which distinguish cat hairs from dog hairs, is the root shape. Dog hairs bear spade shaped roots resembling a closed umbrella; cat hairs bear an indistinct fibrillar root (Hicks, 1977). For the inexperienced examiner these general characteristic root shapes are easy to observe and in this author's opinion the forensic community generally accepts that an animal hair bearing a spade shaped root indicates that it has originated from a dog and not from a cat.

However, during the course of the investigation a few animal hairs although bearing a spade shaped root, the scale cast patterns and the medullary indices indicated cat, not dog, in origin. These roots were considerably shorter than the spade shaped roots seen on the hairs identified as dog in origin. This finding prompted this author to conduct a preliminary study to determine whether cat hairs do possess shaped roots and if so, to measure the lengths.

Dogs	Cats
Blue heeler cross	Siamese
Bull terrier	Blue Burmese
Samoyed	Ginger
Coolie	Tonkanese
King Charles Spaniel	Ginger/white
Golden Retriever	Black/white
Boxer	Ginger
Beagle	White
Chihuahua	White
German Shepherd	Black/white
Pomeranian	Tabby
Red heeler	Tabby
Rottweiler/Kelpie cross	Cream Burmese
German Shepherd/Labrador cross	Black/white/gray
Fox Terrier	Ginger/white
Kelpie	Cream Burmese
Alaskan Malamute	Tabby
Alaskan Malamute	Tabby
German Shepherd/Collie cross	Ginger
Border Colie cross	Black/white
German Shepherd	

Figure 1.13
List of the 41 cats and dogs whose roots were examined.

Guard hairs from 21 dogs and 20 cats (see Figure 1.13) were collected by grooming an animal, either by hand or with a brush, along the entire length of its back. Some smaller breeds of dogs were deliberately chosen in addition to larger ones in order to determine if the spade root characteristic lengths varied with the size of breed.

Guard hairs were chosen because these hairs exhibit characteristics that are the most useful in identifying the animal of possible origin. No attempt was made to differentiate between hairs from different parts of the body, as the forensic scientist is not usually concerned from which part of the body a particular hair originated.

Ten hairs were selected at random from each sample, resulting in a total of 411 hairs (199 hairs from cats and 212 from dogs). The unequal number of hairs for each animal was due to some of the hairs not bearing roots. Each of the 10 hairs was individually mounted in XAM mounting medium, on labeled microscope slides. The roots were examined on a compound, transmitted light microscope capable of up to 400× magnification. The shapes of each root and their lengths were noted. The lengths were determined in microns using an eyepiece graticule calibrated according to the manufacturer's instructions.

The examination of the cat hairs revealed that approximately 30% of the hairs examined exhibited a spade root comparable to the ones seen on the dog hairs. The spade-shaped cat hair roots were not limited to any particular breed of cat. The most significant difference between the spade roots of the two

Figure 1.14
Scanning electron micrographs of cat and dog guard hair roots exhibiting spade shaped roots of differing lengths.

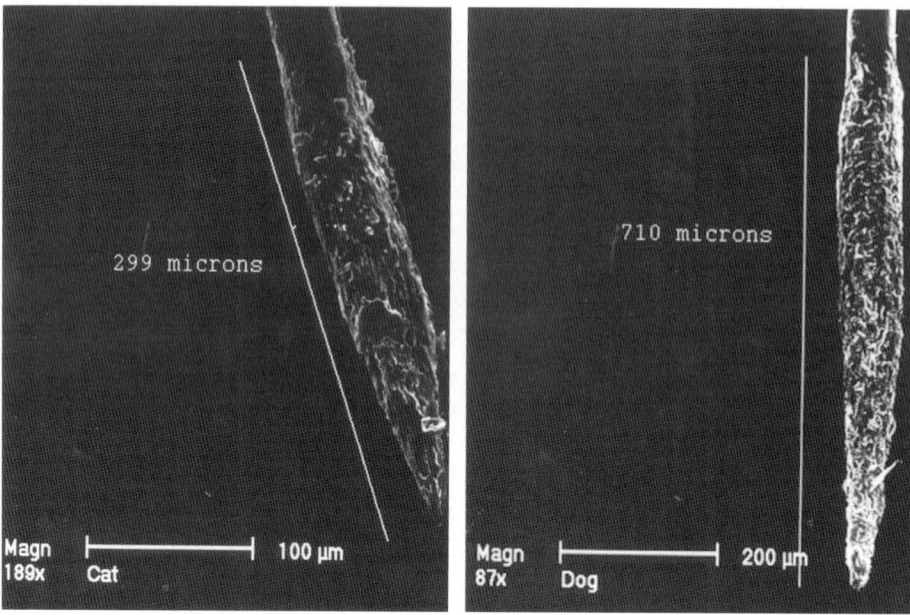

Figure 1.15
Distribution of root lengths (u) for cats and dogs.

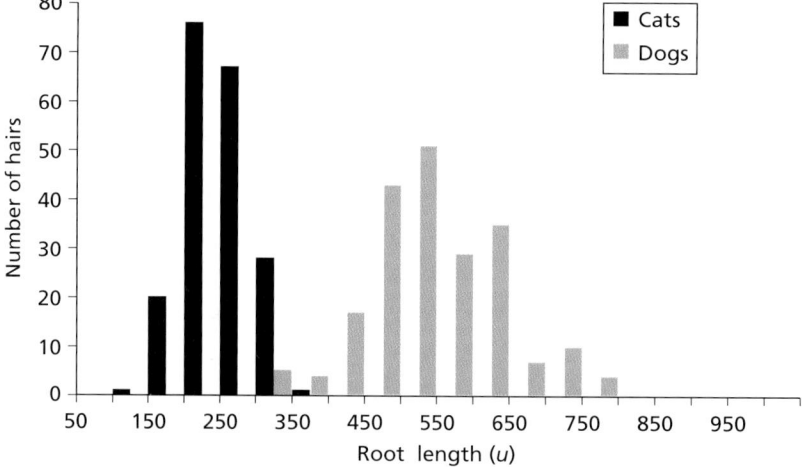

genera was length (see Figures 1.14 and 1.15). The average length of the cat spade roots was approximately 200 μm, while those of the dog spade roots were approximately 500 μm.

The results of this preliminary study highlighted that while the standard works (Appleyard, 1978; Brunner and Coman, 1974) serve as excellent guidelines to the identification of animal hairs it is crucial that the hair examiner, experienced or otherwise, to be aware that they are not definitive or exhaustive works. The results of the study indicate that if the determination on whether a putative hair is of cat or dog origin is based solely on the appearance of a spade root an erroneous identification may be made.

ACKNOWLEDGEMENTS

The author would like to thank Aggie Janusz for her diligence and commitment in conducting the research aspect of the work and Dr. Hilton Kobus for his patience and advice. The author would also like to thank the RSPCA, animal shelters, the South Australian Dog Breeders Association and D.O.G.S. and owners for their co-operation and patience in allowing me to inspect and collect samples from their dogs. Gratitude goes to Dr. Marilyn Henderson, CEMMSA, South Australia, for her skills in electron microscopy and to Reed Education, Australia, for kindly granting permission to reproduce illustrations from "The Identification of Mammalian Hairs". Also thanks go to Dr. G. Roe (Forensic Science Service (UK)) for his advice, Dr. Paul Kirkbride, Dr. Anne Coxon, Dr. Douglas Elliot and Sue Vintiner for proof reading this article, their contributions were invaluable. Last, but not least the author is appreciative of the dogs, Ben in particular, for their patience and docile natures during the collection of their hair samples.

REFERENCES

Appleyard, H.M. (1978) *Guide to the Identification of Animal Fibres*. Leeds, England: Wool Industries Research Association.

Brunner, H. and Coman, B. (1974) *The Identification of Mammalian Hairs*. Melbourne, Australia: Inkata Press.

Conan Doyle, A. (1986) *Sir Arthur Conan Doyle's celebrated cases of Sherlock Holmes*. London, England: Octopus Books Limited.

Freckleton, I. and Selby, H. (1993) *The Forensic Examination of Hairs and Fibres*, Chapter 88. Sydney, Australia: The Law Book Company.

Fridez, F., Rochat, S. and Coquoz, R. (1999) *Science & Justice*, 39, 167–171.

Gaudette, B.D. and Tessarolo, A.A. (1987) *Canadian Society of Forensic Science Journal*, 32(5), 1241–1253.

Grieve, M.C., Dunlop, J. and Haddock, P.S. (1989) *Forensic Science International*, 40, 267–277.

Hicks, J. (1977) *Microscopy of Hair – A Practical Guide and Manual – FBI Publication*, Issue 2, Federal Bureau of Investigation, Washington, DC.

Lowrie, C.N. and Jackson, G. (1994) *Forensic Science International*, 64, 73–82.

Menotti-Raymond, M., David, V.A., Claiborne Stephens, J., Lyons, L.A. and O'Brien, S.J. (1997) *Journal of Forensic Sciences*, 42(6), 1039–1051.

Menotti-Raymond, M., David, V.A. and O'Brien, S.J. (1997) *Nature*, 386, 774.

Moore, J.E. (1988) *Journal of the Forensic Science Society*, 28, 335–339.

Peabody, A.J., Oxborough, R.J., Cage, P.E. and Evett, I.W. (1983) *Journal of the Forensic Science Society*, 23, 121–129.

Savolainen, P., Rosen, B., Holmberg, A., Leitner, T., Uhlen, M. and Lundeberg, J. (1997) *Journal of Forensic Sciences*, 42, 593–600.

Savolainen, P., Arvestad, L. and Lundeberg, J. (2000) *Journal of Forensic Sciences*, 45, 990–999.

Wildman, A.B. (1954) *The Microscopy of Animal Textile Fibres*. Leeds, England: Woolen Industries Research Association.

CHAPTER 2

FIBER-PLASTIC FUSIONS AND RELATED TRACE MATERIAL IN TRAFFIC ACCIDENT INVESTIGATION

Georg Jochem
Forensic Science Institute, Fiber section, German Federal Police Office
Wiesbaden, Germany

INTRODUCTION

Imagine the situation after a severe road crash when a fully occupied car has left the road, struck a tree, and overturned. Two of the occupants may have been ejected from the car and killed, while the other two people may have been found lying seriously injured inside the wreck. The whole place is full of wreckage and busy rescue people are trying to save the lives of the injured ones and to get them out of the car. In such a situation, recovery and life saving have absolute priority to anything else; therefore, in that moment probably no one will ask who has been the driver of the car. But from the investigator's point of view this question can arise very soon and, in many cases, it might be quite difficult to find a satisfying answer.

The seating arrangement may not be determinable through testimonial evidence from eye witnesses of the accident – if there are any – or the rescue people. Quite frequently, the surviving occupants themselves have lost their memories of the accident because of the violence of the event. Furthermore a person who in fact drove the car may attempt to evade responsibility for the accident by claiming that someone else, especially a dead occupant, was driving.

Therefore, the investigator will need physical evidence, which will give him a substantial reflection of the situation in the car at the moment of the accident. He must look for "traces" that have been caused by the accident, the examination of which leads to clear statements in the testimony of the expert witness. From this train of thought, the question arises which kind of evidence will be helpful at all.

First of all, when discussing this question one must keep in mind that, in most cases, people involved in accidents either are driving in their own car or are occupants in the car of their family or a friend. Therefore, fibers or hairs on the seats commonly collected with adhesive tape have only very low, if any, evidential value in these cases: this is because it is almost impossible to determine the exact time of the transfer of such loose particles. The occupants may have changed places a short time before or after the accident or may have left or been pulled out the car by the rescue service through the same door. In fact, it has been possible to draw useful conclusions from the fiber distribution, for example between the driver's and the front passenger's seat, but these cases are very rare. If the occupants did not wear their seat belts and the car was involved in a head-on collision, the investigator may find hairs or tissue pinched in the cracked windscreen probably leading him to the person who sat behind where the cracks occurred.

The examination of the bloodstains also may give clear results and the seating arrangement may be determinable in this way, but for that purpose the occupants must suffer from bleeding injuries and a clear pattern resulting from an impact, such as of a head in the windscreen, must be present. If no pattern or only some small bloodstains on various areas in the vehicle are found, their interpretation may be problematic, even worse, misleading. This is especially true in roll-over situations. In some cases it may also be possible to find patterned injuries on the bodies of the occupants resulting from the seat belts or from an impact on the steering wheel. Perhaps even marks or smears resulting from the pass of the seat belts over their clothing may be present.

Trace material suitable for further examination including fibers, saliva, or blood is quite often present on released airbags. Again this evidence must be assessed very carefully. This is particularly true if the car has overturned or if any modification, for example, by recovery of all occupants through one door or simply by the weather conditions after the accident cannot be excluded.

Fortunately, other trace evidence with a very high evidential value frequently occurs in traffic accidents. Best known among this group of trace evidence are fiber-plastic fusions (FPFs) and plastic coating marks (Bürger, 1977, 1989; Jochem, 2001; Jochem and Pabst, 2000; Krauß et al., 1993; Masakowski et al., 1986; Metter, 1978; Pabst, 1984, 1992; Schiller, 1995; Schwarz et al., 1985). Once formed, these traces are resistant to the above-mentioned factors, which may affect other kinds of evidence. However, the complete destruction of the car (e.g. by fire) or the occupant's clothing (e.g. by themselves, relatives, or the hospital staff) of course would affect or even destroy the trace material required for identification and comparison.

FORMATION OF FPFs AND RELATED TRACES

FPFs AND PLASTIC COATING MARKS

In a collision of a car with an obstacle, trace evidence may be exchanged between the interior surfaces, which are made of thermoplastics (Putnam, 2001), and the clothing of the occupants. According to the direction of the impact, the occupants will be catapulted onto parts of the interior equipment of the vehicle. As a result, parts of garments are frequently rubbed under high pressure against surfaces of thermoplastic components and the kinetic energy is transformed into frictional heat. This causes local melting of the thermoplastic material.

During these contacts, which will last only a split second, textile fibers of the rubbing garment are transferred into the softened plastic and are fixed in the immediately re-solidifying material (see Figures 2.1–2.3). Sometimes even impressions of the fabric, for example of the denim twill lines, are produced, as shown in Figure 2.4.

By comparison of the clothing of all occupants with the transferred fibers, the latter can be assigned to or excluded from a distinct garment. Therefore,

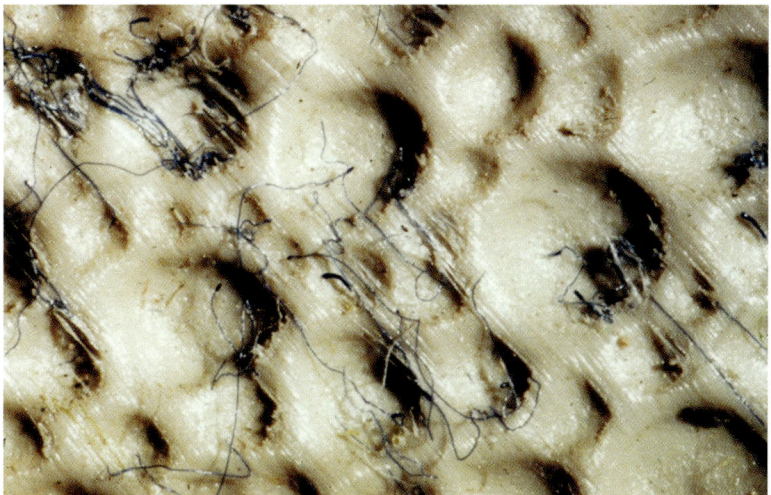

Figure 2.1
FPF on a sun visor with embedded blue polyester fibers.

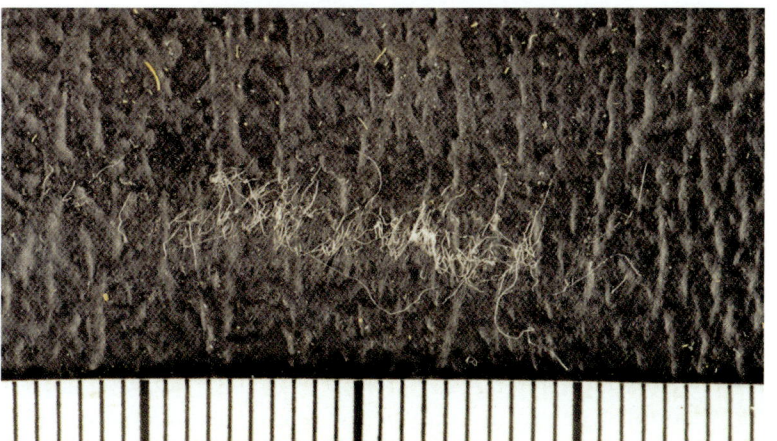

Figure 2.2
FPF on an interior paneling of a driver's door with embedded colorless acrylic fibers.

Figure 2.3
FPF on an interior handle of a driver's door with embedded red cotton and polyester fibers.

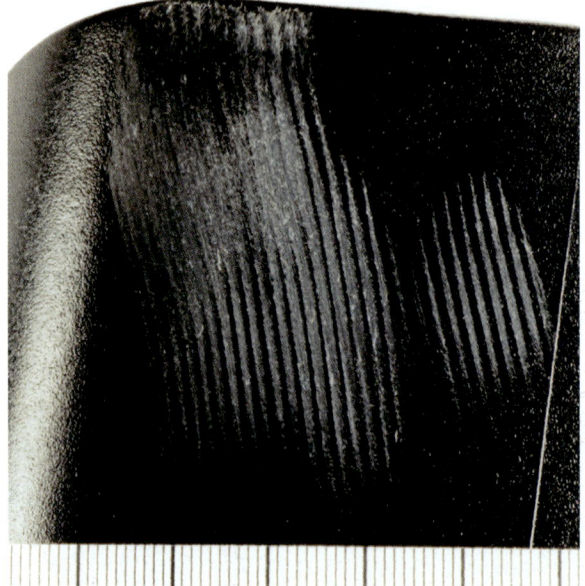

Figure 2.4
FPF on a cover of a steering column with embedded blue denim fibers; note the impression of the denim fabric twill lines.

these FPFs are a kind of snapshot, showing the examiner which garment was in which place at the very moment of the impact.

In addition, in most cases, FPFs show threads of smeared plastic which indicate the direction of the impact that produced them. Hence, the outcome of an FPF examination may not only be the seating arrangement of the occupants of an automobile at the moment of the impact, but may also help to determine the trajectories of the vehicle's occupants during the accident. In some more complicated cases, for example if a car overturns several times and people are ejected from the car, it might be necessary to consult a technical expert who is able to analyze the movements of the occupants or to simulate them by means of a computer program like "PC-Crash" (Turner, 1999).

One of the most often asked questions in connection with FPFs is what velocity will cause these transfers to be formed in an accident. It is impossible to answer with a definite value. Several parameters that influence the formation of plastic fusions, such as the deceleration factor, the angle of impact, the size of the contact area, the surface structure, and the melting point of the plastic material, are more important than the velocity of the car just before the crash. As a general rule, the formation of FPFs can be expected if the car's body is severely damaged. The use of seat belts and the release of airbags will lower the probability of the formation of such trace material, but if the impact has a lateral component (side impact), FPFs will probably be formed.

However, it must be emphasized that under normal driving conditions no FPF or other similar trace material described in this chapter will be produced in a car – even if somebody jams hard on the brakes. The only known exceptions from that rule are plastics with a very high content of softening agents, which may be found in a few isolated cases in the soles of shoes or as artificial leather used for inside covers of doors and jackets. If such a material is involved, the examiner must assess the trace material carefully.

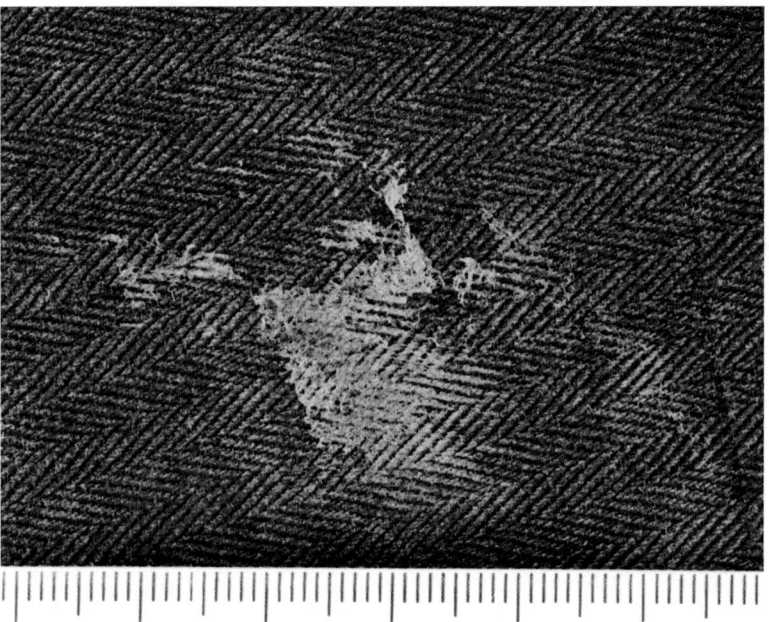

Figure 2.5
Plastic coating mark on a blue jacket.

58 TRACE EVIDENCE ANALYSIS

The corresponding traces to FPFs on a car's interior surfaces are plastic coating marks on the clothing of the occupants. These traces are formed somewhat less frequently – usually in high-speed impacts – by transfer of softened thermoplastic material from the interior of the car to the garments (see Figures 2.5–2.7). They are rather valuable, particularly if garments showing a low sheddability are involved (e.g. leather or filament yarn fabric) or if two occupants of a car are wearing garments made of indistinguishable fibers (such as blue denim jeans or white cotton shirts).

An exchange of plastic coating marks may also happen between the shoes of the driver and the pedals. In most of these cases, the marks will be found on the

Figure 2.6
Large plastic coating mark on a trouser.

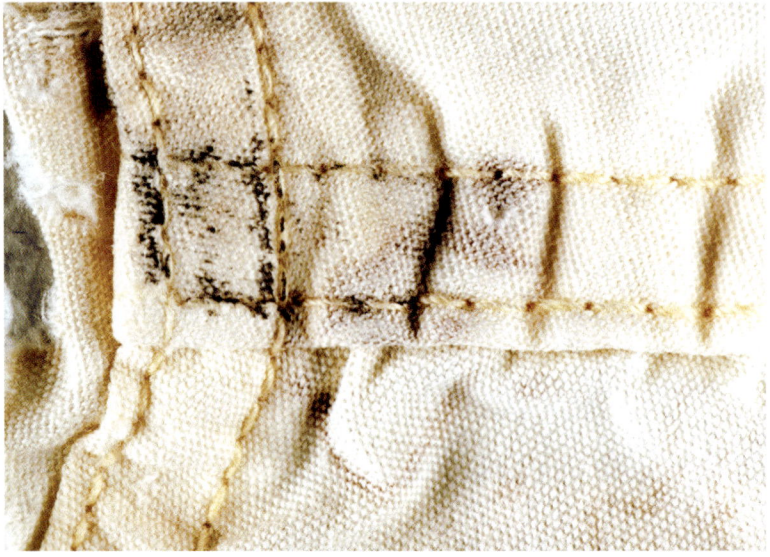

Figure 2.7
Small plastic coating mark on a shirt's shoulder seem.

brake or the clutch pedal; the gas pedal is inevitably less affected (see Figures 2.8 and 2.9). The pattern of the pedal may be pressed into the sole of the driver's shoe – or vice versa – as shown in Figure 2.10 a pattern of a profiled sole may be found on the pedal (Bürger, 1977b; Lautenbach and Schaidt, 1970).

Figure 2.8
Plastic coating mark on the clutch of a break pedal from the sole of the driver's shoe.

Figure 2.9
Plastic coating mark from the clutch of a pedal on the sole of a driver's right shoe.

Figure 2.10

Pattern of the profiled sole of the driver's shoe on a clutch of a break pedal.

Figure 2.11

Transfer of a textile printing's material from a T-shirt to a steering wheel.

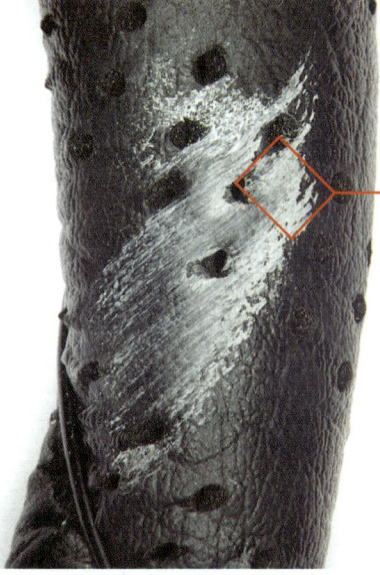

TEXTILE PRINTINGS AND LEATHER VARNISH

Quite rarely seen are fusion marks caused by applications or coatings on the surface of garments. On T-shirts or pullovers one may find printed material like a logo or a picture. This material, typically acrylic resin, may be transferred together with the fibers. Figure 2.11 shows one example for such a transfer from a printed T-shirt to the cover of a steering wheel. Leather clothing may be colored or made weather proof by a water repellent finish. During accidents these materials, typically an acrylic resin or polyurethane, may also be transferred from the clothing. Figure 2.12 shows an example for such a transfer from a leather jacket to the cover of an A-pillar.

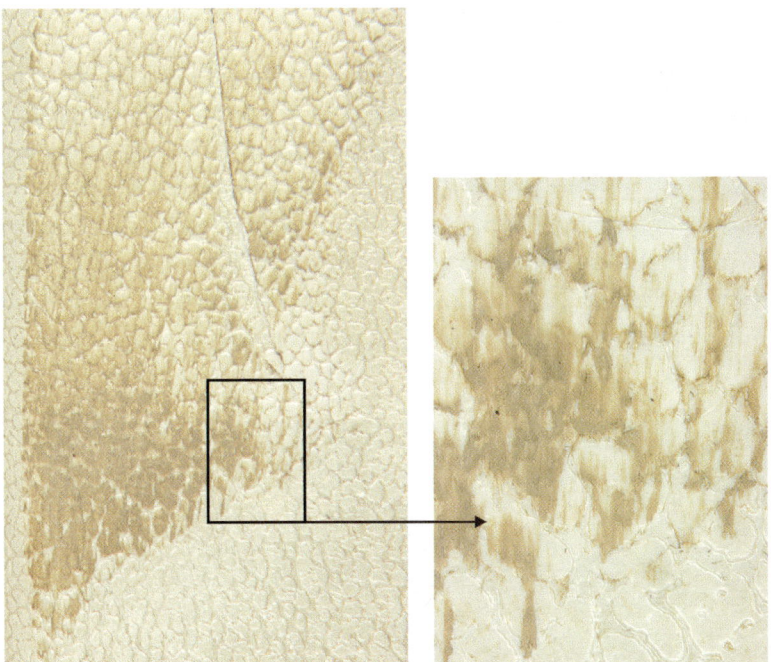

Figure 2.12
Transfer of leather varnish from a leather jacket to the cover of an A-pillar.

Figure 2.13
Fusion mark with embedded hairs on a roof lining.

HAIRS

If somebody hits a thermoplastic part of the car's interior with his head, some hairs may be transferred (just like fibers) into the molten plastic (Masakowski *et al.*, 1986). Of course, such trace material will be found predominantly at the upper parts like the pillars or the roof lining (see Figure 2.13).

PINCHED FIBERS

If a passenger hits a part of the interior equipment of a car almost or exactly perpendicularly, no frictional heat will be produced to melt the thermoplastic material but it will crack from the sudden deformation. Fibers can be pressed into the cracks and become pinched in them after recovery of the previous shape of the plastic component. Pinched fibers are recognized sometimes, mostly on padded parts like the dashboard (Pabst, 1984, 1992; Schwarz *et al.*, 1985). Examples are given in Figures 2.14 and 2.15; note that the cracks in the plastic material may only show up by the fibers sticking out from the surface.

Figure 2.14
Pinched blue denim cotton fibers in a cracked dashboard.

Figure 2.15
Pinched blue denim cotton fibers in a steering column cover.

SEARCHING FOR TRACE EVIDENCE

It is essential for a successful examination that the outerwear of all occupants, including socks and shoes, is secured as soon as possible after the accident, in particular the clothing of severely injured or dead persons. The evidentiary value of the clothing may not be recognized by emergency or hospital personnel, who may destroy or throw away the garments.

The examination of the car's interior should be carried out in a large, well-lit, and dry area. Direct sunlight and – even worse – wet surfaces will hinder the examiner from finding the evidence. To locate FPFs in the car, it takes experience, a keen eye, and proper lightning (Putnam, 2001). During the search process it is essential that a light is used which produces a white, regular, and diffuse light, such as a neon lamp. The use of a flashlight is not recommended because it illuminates the search area too irregularly. The examiner should use a combination of oblique and direct lightning as well while inspecting all plastic surfaces in the car which could bear FPFs; the use of a handheld magnifying glass (2–10× magnification) during this inspection is highly recommended (Putnam, 2001).

It is important that the car is examined in order to find any trace material that may give information concerning the seating position of each occupant at the moment of the accident. The examiner, therefore, must not search only for trace material caused by the driver. What is more, he must search the whole car, because sometimes it is possible to determine the driver indirectly by exclusion of all the other occupants (see Case example 1). To find the relevant trace material more quickly it will help the examiner to have an idea of what might have happened in the car during the accident. As stated above, the occupants will be forced to move according to the direction of the impact. For example, if the point of impact is on the left-hand side of the vehicle, the occupants following their inertia will be forced to move (in a first approximation) towards the left-hand side. Subsequent veering and overturning of the vehicle may complicate the situation. But in general, it always makes sense to look first for trace material on plastic parts at the side of the first impact because most of the kinetic energy of the car and its occupants will be taken away at this instant.

As Putnam (2001) states, the examiner should inspect surfaces for any change in the texture. If there are any changes, especially if the surface appears abraded or scratched, this area should be inspected further for fibers embedded in the plastic, fabric impressions, etc. Cracked parts should be checked for pinched fibers. Sometimes the examiner may find FPFs quite easily because they are so large or because there is a clear contrast in color or brightness between the embedded fibers and the surface. Unfortunately, in many cases these traces are quite small and the fiber fragments may not show up clearly even by looking through a good magnifying glass, but can be easily seen while examining the surface with a stereomicroscope. Areas where a first inspection cannot rule out the presence of an FPF or other relevant trace material should be submitted for further examination in the laboratory. Nevertheless, everything should be documented, photographed, and the panel (possibly) bearing an FPF should be secured or the relevant area cut out (Putnam, 2001). It is absolutely necessary that the evidence is not collected by taping or picking the fibers from the surface; otherwise a lot of information concerning structure, context, and direction of the fusion mark will be irretrievably lost.

There are many parts of the interior equipment on which the examiner may find FPFs and related trace material including dashboard, door covers, steering wheel, and pedals, respectively (see Figures 2.16–2.18).

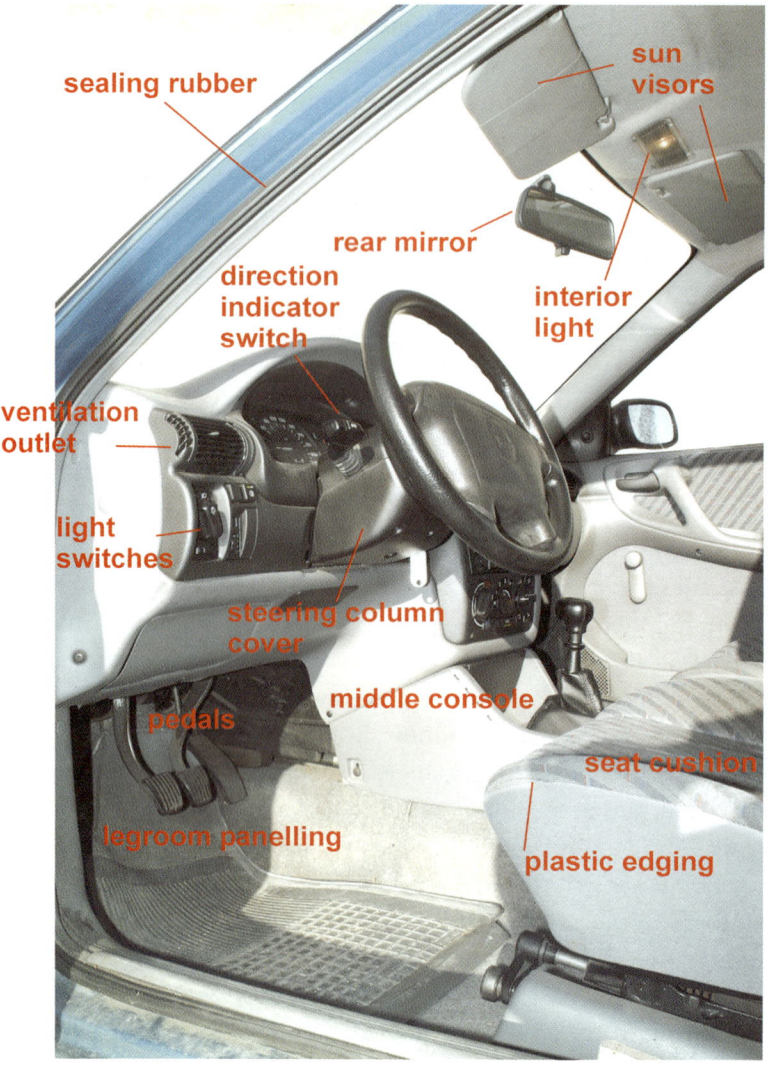

Figure 2.16
View into a car from the driver's door showing some of the plastic parts which are regularly checked for FPFs.

Contrary to a widely held view, FPFs are quite rarely found on seat belts. The reasons for this are the tightness and, in comparison to other plastic parts in the car, the high-melting point of the polyester filament yarn used for the production of seat belts. On the other hand, the achievable frictional heat due to the normally short rubbing between the seat belts and the clothing is very low. Therefore, FPFs on seat belts are likely to be found only in very high-speed impacts.

Figure 2.17
View on a car's dashboard showing some of the plastic parts which are regularly checked for FPFs.

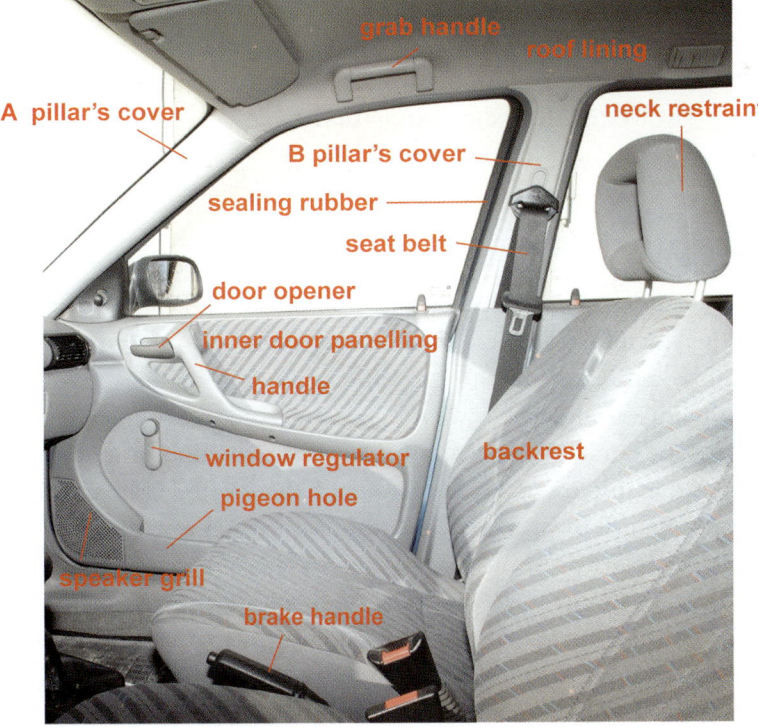

Figure 2.18
View into a car showing some of the parts which are regularly checked for FPFs.

Depending on what happened during the accident, some traces may be more useful than others to determine the seating arrangement. For example, in the most "simple" head-on crashes every FPF or related trace material in the surroundings of the driver's seat will be assigned to the driver. This may include trace material possibly found on the pedals, the cover of the steering column, the steering wheel, the left A- and B-pillars, the left side of the dashboard, the middle console and the roof lining, and traces on parts on the inside of the driver's door. But if the course of the accident is more complicated, the assignment of trace material on some of these areas to the driver might become debatable, because it might be difficult to rule out that other occupants might have caused it as well. This is especially true if the car has overturned several times and if the occupants did not wear seat belts. Then they might have had enough room to move and to reach many places in the car, may be they even will have been catapulted out of the car. In such a case trace material in the region of the lower limbs will be of special interest, because only this can be assigned unequivocally. In the surroundings of the driver's seat, this may be found on and around the pedals, the lower part of the dashboard, and the cover of the steering wheel column; no matter how the other occupants will have moved, they would not have contacted the legroom of the driver (see Case examples 3 and 4).

These considerations underline that the forensic scientist needs as much information as possible concerning the course of the accident from the traffic accident investigator and the position of the trace material in the car from the crime scene personnel in order to assess the case completely. Furthermore, experience from routine casework has shown that it is even better if the scientist is involved in the examination of the car.

EXAMINATION OF TRACE MATERIAL

In the laboratory, the evidentiary automotive parts and garments are searched further for FPF evidence. In a first step, they are placed in a well-lit area and examined thoroughly with the aid of a magnifying lens, followed by an inspection of areas of interest under a stereomicroscope. If bigger parts or large areas must be examined, the use of a stereomicroscope mounted on a swivel arm (operation microscope) is recommended. As stated above, FPFs may be large or eye-catching, but in many cases these traces tend to be very small. The same applies to plastic coating marks on garments. Of course they will generally have a glossy appearance, but, however, their color may be so similar to the garment's color or they can be so small (only a few threads or even fibers of the garment may be affected) that it will be difficult to find them. After documentation of the location and appearance of the trace material it can be collected for further examination. This should be done by picking the plastic material adhering to the garments and the fibers embedded in plastic surfaces from the substrate with the aid of tweezers and a stereomicroscope (Pabst, 1992; Putnam, 2001).

The transferred fibers are examined and compared with the clothing of the occupants using established methods like microscopy, microspectrophotometry, and Fourier transform infrared (FTIR) spectroscopy, described in detail by Robertson and Grieve (1999). Microscopic methods and FTIR spectroscopy are used for the examination of the transferred plastic materials. The latter has been widely used to characterize plastic materials in forensic casework, enabling the examiner to identify at least the basic polymer type, in many cases also copolymers and some additives. FTIR spectroscopy, therefore, is an excellent tool for the comparison of trace material found on a garment with standard samples taken from the car's interior (Putnam, 2001; and references therein).

To enhance the discrimination power significantly, the plastic materials can be compared microscopically. For that purpose the questioned and standard material have to be transformed into thin films, which are produced by heating the samples to about 120–150°C and pressing them to thin films of 5–10 μm (see Figure 2.19) with the aid of a small press which was especially constructed by Pabst (1984, 1992). Even very small amounts down to some nanograms of the plastic material will be sufficient for an examination and are manageable with a steady hand.

As shown in Figure 2.19, the obtained polymer films of trace and standard material are compared using microscopical methods such as bright field, dark field, and polarized light. In the chosen example, no difference between the questioned and standard material concerning the filler components and the color can be observed under any kind of illumination. Such a perfect match is quite unlikely observed just by chance, because samples taken from similarly colored plastic parts of one car regularly reveal a large variety concerning their color and their filler components to the examiner while looking through his microscope (Pabst, 1984, 1992). This is clearly demonstrated in Figure 2.20, showing pictures of polymer films taken at bright field, dark field, and polarized light, respectively, which have been obtained as described above from seven different black plastic components of one car.

The impressive variety in the chemical composition and also in the optical properties of the plastic materials in a car is a direct result of the manufacturer's efforts to enhance the mechanical and aesthetic properties of these parts. For functional reasons automotive engineers will have to place chemically different materials in different parts of the car which may be in close proximity to each other, but for aesthetic reasons they attempt to make them visually similar (Pabst, 1992; Putnam, 2001).

Figure 2.19
Examination scheme for transferred plastic material and comparison of the obtained films at bright field, dark field, and polarized light (top to bottom).

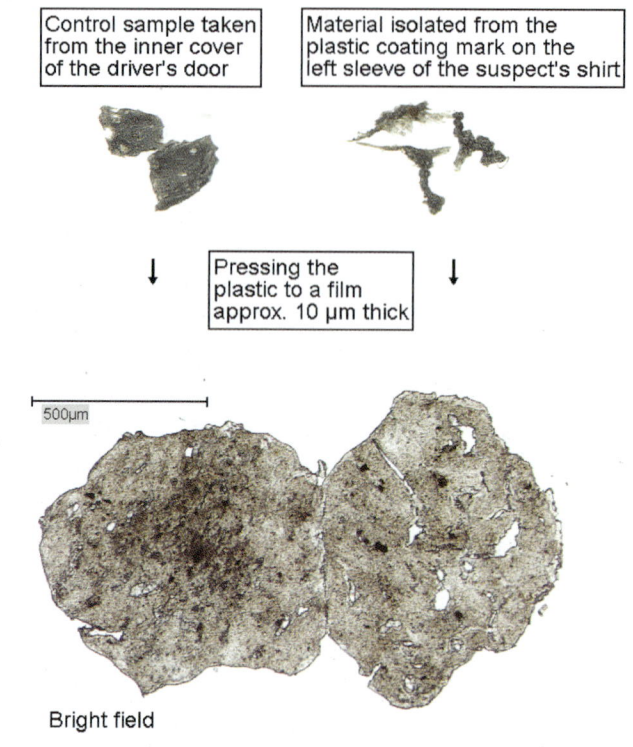

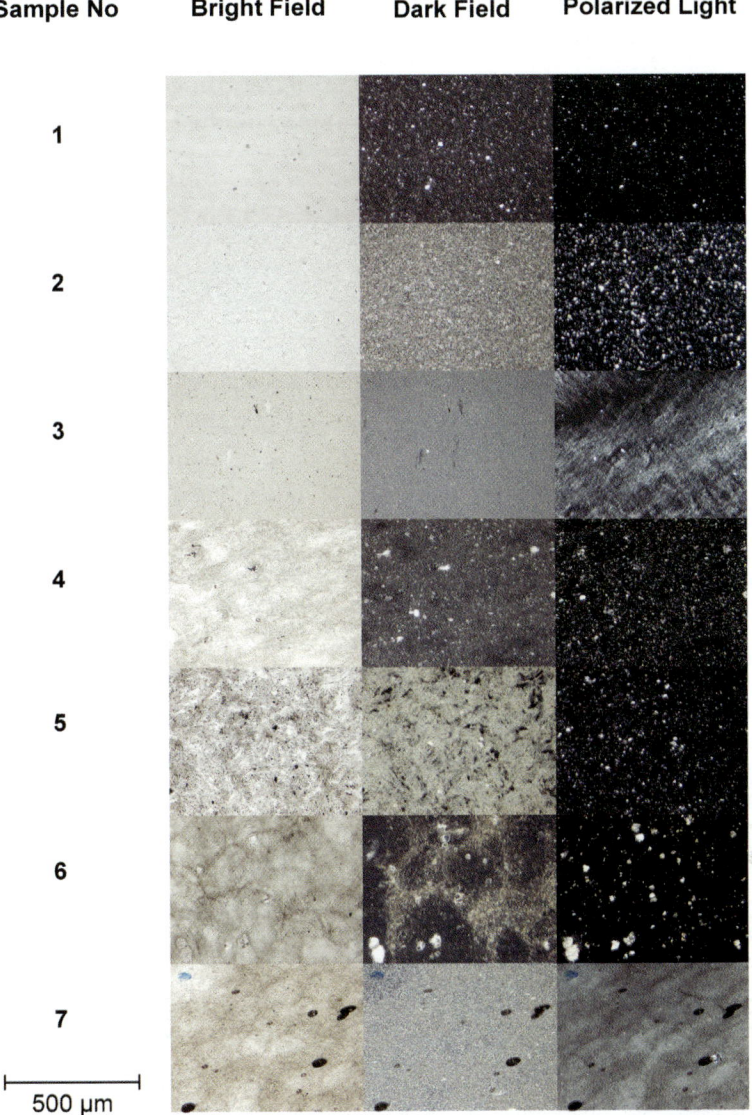

Figure 2.20
Microscopic pictures taken from polymer films at bright field, dark field, and polarized light (left to right column). The films were obtained as shown in Figure 2.19 from seven different black plastic samples taken from the interior equipment of one car.

EVIDENTIAL VALUE

The traces described above have a very high evidential value, because

- they are formed exclusively during traffic accidents and
- in the "closed system" of a car only
    - a limited number of occupants with
    - a limited number of items of clothing will have contact with
    - a limited number of different interior components made of thermoplastic material.

This fact enables us, in most cases, to come to unequivocal assignments and clear statements just by the analysis of the interaction of mass products like fibers and plastics without the use of DNA or fingerprints. The case examples described in this chapter clearly demonstrate the value of this type of evidence.

RECONSTRUCTION OF THE SEATING ARRANGEMENT IN VEHICLES

CASE EXAMPLE 1

A car occupied by four males veered off a road and struck a tree on the right side of the road. The point of impact was at the right front wheel (see Figure 2.21). Two men were seriously injured and one occupant died. The fourth one – the owner of the car who was suspected of having been the driver – was only slightly injured. The two seriously injured occupants were unable to tell the police what happened and the suspect gave no statement.

Examination of the car revealed several FPFs which must have been caused by the front seat and the back seat passengers, but no FPFs were found which may have resulted from the driver's clothing. Nevertheless, after examination of the parts of the interior equipment of the car and the clothing of the occupants in the laboratory, the seating positions of the killed and the seriously injured persons were determined. On the right side of the dashboard, large FPFs with embedded blue denim cotton fibers were found (see Figure 2.22). Only one of

Figure 2.21
Wreck of the car showing the point of impact at the right front wheel (Case 1).

the occupants (person B) wore such a pair of blue jeans. Examination of the jeans revealed transferred plastic material in the area of the right knee matching the polyvinylchloride material of the dashboard. A cross-transfer of trace material, therefore, proved this contact.

Two more FPFs were observed on the interior paneling of the right front door. Once again blue denim cotton fibers matching only the fibers in the blue jeans of person B were embedded in one of them. In the other one, dark blue cotton fibers were found. These fibers only matched those in the T-shirt of person B.

FPFs were present on the regulating wheels of the backrests of the front seats, which, in this case, were placed at the side of the backrests facing the middle of the car. On the regulating wheel of the front passenger's seat we even recognized a textile pattern (see Figure 2.23). The embedded dark gray cotton fibers and the textile pattern were assigned to the black jeans of person C. The corresponding

Figure 2.22
The highly deformed right part of the dashboard (Case 1); the red frames mark the FPFs with embedded fibers from the jeans of person B. Although this person had fastened his safety belt he hit the dashboard with his knees because of the strong deformation of the car in that area.

Figure 2.23
Regulating wheel of the front passenger's backrest showing the FPF with a textile pattern; the corresponding plastic coating mark was found on the black jeans of person C (Case 1).

plastic coating mark, resulting from the intense contact with the backrest regulating wheel and made from an acrylonitrile–butadiene–styrene copolymer (ABS), was found in the area of the right thigh of these jeans (see Figure 2.23), thus proving the contact as a cross-transfer of trace material. According to the direction of the impact at the right front wheel of the car these traces indicated that person C could have sat only on the left side in the back of the car.

On the rear side of the right front seat's backrest a fusion mark including dark gray polyester and viscose fibers was found. These fibers only matched those in the trousers of the dead person D, who, therefore, must have sat on the right side in the back of the car.

Figure 2.24 shows the resulting assignment of the FPFs to the seating positions and the textiles worn by the occupants. According to these results the persons B, C, and D were named as front and back seat passengers, respectively, and person A was the only possible driver. As shown in this case, it is possible to determine the responsible driver indirectly by exclusion of all the other occupants. However, in order to do this, and as stated above, the examiner will

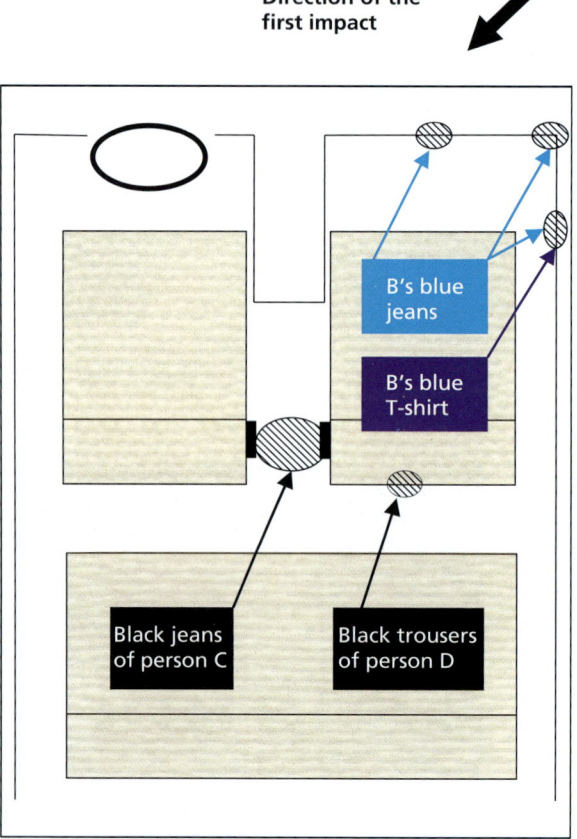

Figure 2.24
Scheme of the car's interior showing the assignment of the FPFs to seating positions and clothing (Case 1).

need the complete outerwear of every involved person and he should not only search for traces caused by the driver but he will have to examine the whole car.

CASE EXAMPLE 2

A VW Transporter occupied by two males veered off a narrow road touching the crash barrier on the right side of the road. The vehicle then veered back on the road where it collided head-on with an oncoming car (see Figure 2.25). The occupants of the Transporter and the woman driving the car were seriously injured. The seating arrangement of the Transporter could not be determined through testimonial evidence because the ambulance men could not tell the police who sat in the driver's seat and the occupants themselves lost their memories of the accident.

The clothing of the Transporter's occupants was seized and we examined the interior of the Transporter. On the cover of the steering column, an FPF with embedded blue denim cotton fibers running from the left to the right was found. This trace must have been caused by the driver's trousers. Unfortunately, both occupants wore light blue jeans, so assigning the transferred fibers to one of them solely by examination of the transferred fibers was impossible. However, examination of both pairs of trousers revealed a corresponding plastic coating mark running from the right to the left in the area of the left knee of the jeans

Figure 2.25
Wreck of the VW Transporter showing the point of impact at the front (Case 2).

of person A and the transferred material matched that of the steering column cover (see Figure 2.26). The contact of person A's jeans with the steering column cover was proven by cross-transfer of trace material.

In addition, an FPF was present on the sealing rubber of the front passenger's door beside the safety belt (see Figure 2.27), caused by the clothing of the passenger, most probably during his rejection after the first impact. The embedded

Figure 2.26

The steering column cover of the VW Transporter (turned upside down, center) showing an FPF with embedded blue denim fibers matching those in the light blue jeans of persons A (left) and B (right) as well; however, the corresponding plastic coating mark resulting from the contact with the steering column cover was found on person A's jeans in the area of the left knee (Case 2).

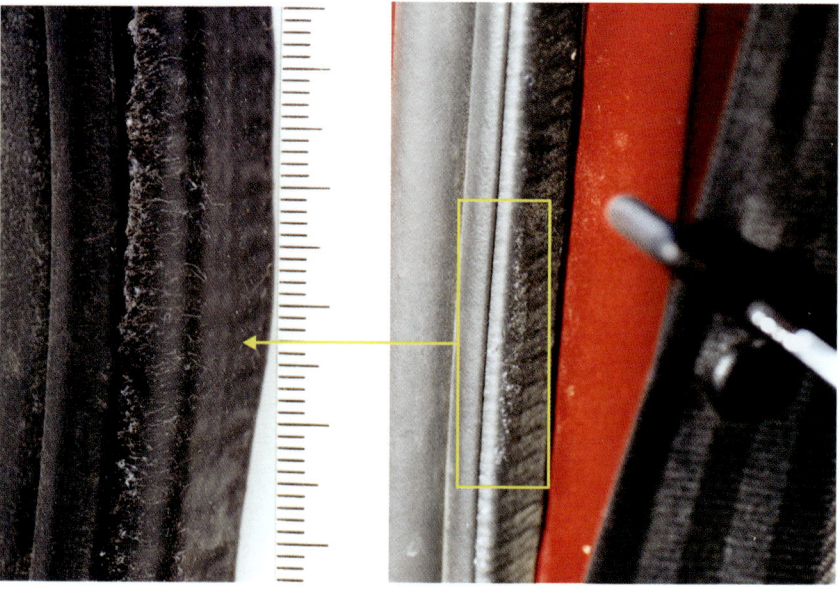

Figure 2.27

FPF on the sealing rubber of the front passenger's door (Case 2).

light violet cotton and polyester fibers matched only the fibers in the sweatshirt of person B. On the backside of that sweatshirt, a plastic coating mark was found; the transferred PVC based polymer matched the material of the sealing rubber.

These results showed that person A had been the driver of the Transporter. Without the plastic coating mark on his jeans we would have been unable to distinguish between the two similar jeans and to determine the driver directly.

CASE EXAMPLE 3

This case involved an imported car occupied by three young males. While driving on a narrow road at a speed of approximately 120 km (75 miles) per hour, the driver lost control of the vehicle in a right-hand curve and the car collided with a tree on the right side of the road. The point of impact was on the left side of the vehicle, immediately in front of the A-pillar. The car broke up into two parts along a line between the two A-pillars. The front part of the car, including the motor, moved to the left side of the street while the remaining part with the occupants rotated around the tree and overturned. Two people were ejected from the vehicle; one of them was found dead and the other one seriously injured just beside the wreck. The driver's seat and the steering wheel were also ejected. The third occupant was found lying dead on the back seat.

Examination of the wreck and the clothing of the persons involved revealed several FPFs. On the cover of the steering column a fusion mark with embedded dark gray cotton fibers was found. These fibers only matched those in the trousers of person A. Examination of these trousers revealed transferred plastic material in the area of the right knee matching the polypropylene cover of the steering column (see Figure 2.28).

The plastic caps of the clutch and the break pedal showed some melting marks. Corresponding marks to these plastic coating marks on the sole of the right shoe of person A were observed. The transferred material on the sole matched the PVC material of the pedals. On the brake handle, the middle of the dashboard, and the hinged lid of the glove compartment, fusion marks with embedded colorless and blue cotton fibers only matching the fibers in the blue jeans of person B were found. On the inside paneling of the right front door a fusion mark, including a large number of dark blue cotton fibers, was observed. These fibers only matched those in the socks of person B.

On the interior light and on the backrest of the front passenger's seat, fusion marks including light blue cotton fibers were found. These fibers only matched those in the T-shirt of person B. Corresponding to the fusion mark on the backrest, a plastic coating mark on the back of the T-shirt of person B was discovered. The transferred material matched the molten polyester fibers of the seat cover (see Figure 2.29). On the upper part of the inside cover of the left back

76 TRACE EVIDENCE ANALYSIS

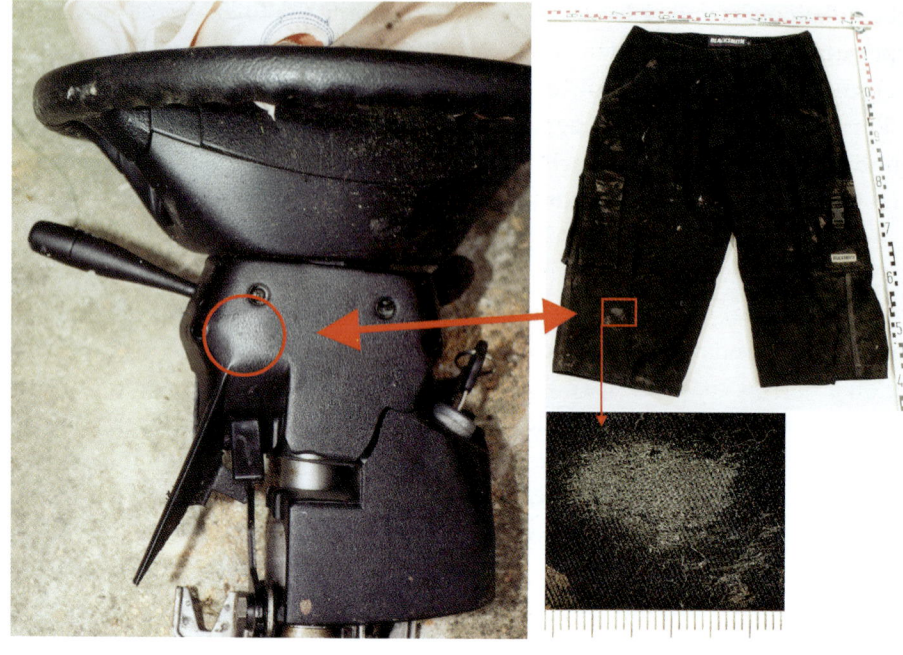

Figure 2.28
The cover of the steering column showing an FPF with embedded black cotton fibers matching those in the trousers of person A; the corresponding plastic coating mark was found on these trousers in the area of the right knee (Case 3).

Figure 2.29
FPF running from the front to the back with embedded light blue cotton fibers on the back rest of the front passenger's seat and the corresponding plastic coating mark on the back of person B's blue T-shirt (Case 3).

door a fusion mark including a large number of light violet cotton fibers was observed. These fibers only matched those in the socks of person C. On the left neck restraint and just below that on the backrest of the back passenger's seat, fusion marks, including colorless cotton fibers and dark gray viscose fibers, were found. These fibers only matched those in the T-shirt of person C. Figure 2.30 shows the resulting assignment of the FPFs to the sitting positions and the textiles worn by the occupants. According to these results, it was obvious that person A was the driver of the car and persons B and C were the front passenger and the back seat passenger, respectively.

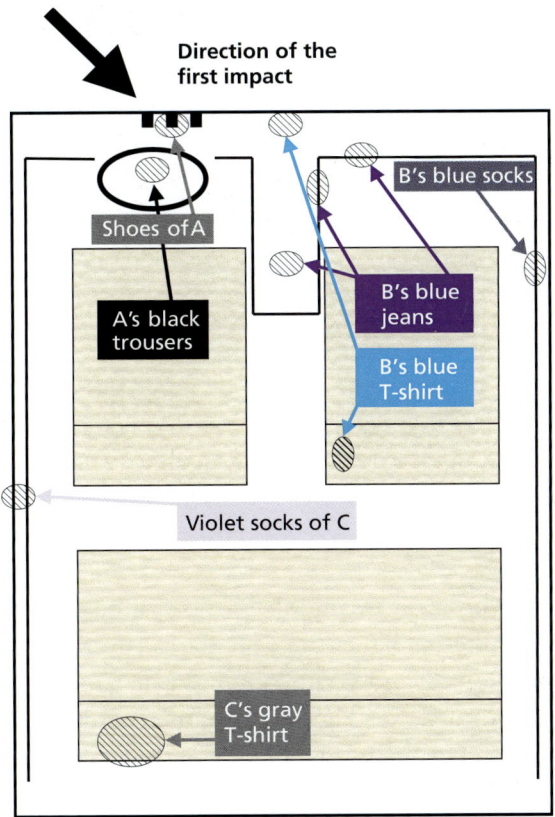

Figure 2.30
Scheme of the car's interior showing the assignment of the FPFs to seating positions and clothing (Case 3).

But at the scene person B was found lying dead on the back seat and person C lay beside the wreck. A mix-up of the clothing of B and C was clearly ruled out; the possible movements of the front seat passenger during the accident had to be considered. According to the examination, none of the occupants was wearing his seat belt. On reconstruction of the accident, a technical expert came to the conclusion that the front seat passenger may have indeed been thrown from his seat into the back of the car. This activity is documented by the direction of the FPFs as shown below.

78 TRACE EVIDENCE ANALYSIS

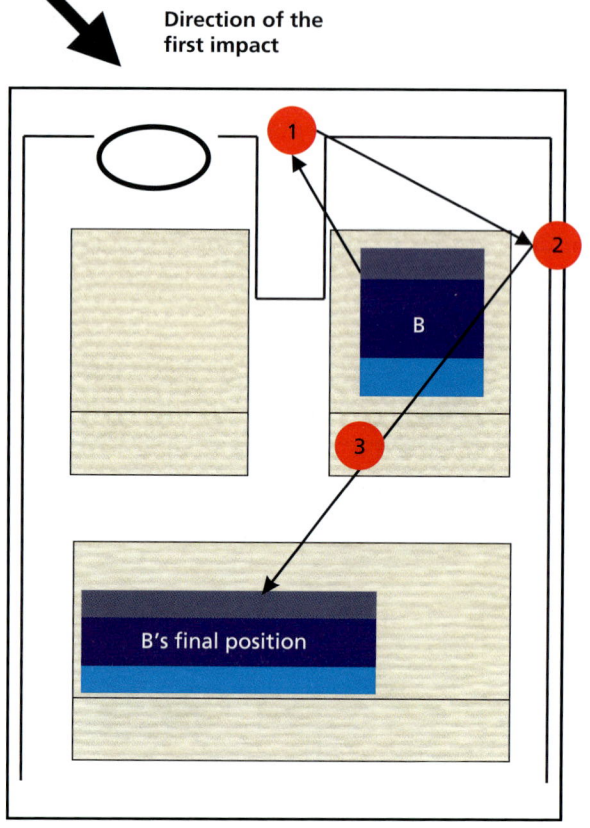

Figure 2.31

Scheme of the car's interior showing the movement of the front seat passenger during the accident (Case 3).

In position ❶ (middle of the dashboard, brake handle, glove compartment, and interior light) the FPF marks run from the back to the front, whereas in positions ❷ and ❸ the FPFs indicate a movement of the textiles from the front to the back (see Figure 2.31). In accordance with these findings person B was catapulted against the middle of the dashboard ❶ upon the first impact, hitting the brake handle, the glove compartment, and the interior light. While the car was spinning and overturning his feet hit the inside paneling of the right front door ❷ and later on he was thrown over the backrest of his seat ❸ into his final position in the back of the car.

CASE EXAMPLE 4

In this case a VW Golf occupied by two young males veered of a road into the right ditch and was thrown back on the road where it overturned. One of the occupants was ejected from the car and died; the other one – the son of the car's

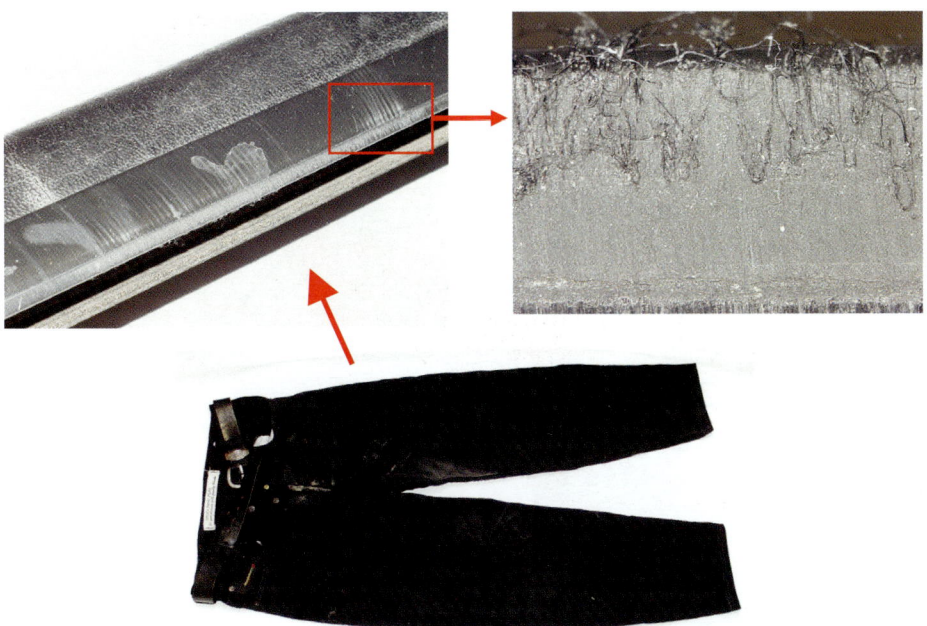

Figure 2.32

FPFs on the interior cover of the driver's door with embedded dark gray cotton fibers matching the fibers of the dead person's black jeans (Case 4).

owner who was suspected to have been the driver – only suffered slight injuries. He told the police that his dead friend has been the driver.

Examination of the car revealed fusions marks which must have been caused by the driver's clothing. Two FPFs were observed at the upper edge of the interior paneling of the driver's door. They were running from the inside to the outside, indicating a movement of the driver through the window of his door. Dark gray cotton fibers matching the fibers of the dead person's black jeans were embedded in both (see Figure 2.32). Two marks of molten plastic running from the bottom to the top were present on the left side of the middle console. At these positions, a thermoplastic material that matched the soles of the dead person's shoes in every respect was found adhering to the nearly unchanged surface of the console. In addition, corresponding melting marks on the right side of his right shoe (see Figure 2.33) were noted. Resulting from these findings it was obvious that the driver had been ejected from the car and that the former suspect had told the truth.

Another interesting part of this case was a really large FPF approximately 60 cm in length on the roof upholstery, the crank-operated sunroof, and the right sun visor. Hundreds of blue denim cotton fibers were embedded in that FPF. The fibers matched those in the blue jeans of the former suspect, who actually must have been the front passenger. The molten parts of the plastic material, formerly beige in color, were now blue colored, changed by abrasion from those indigo

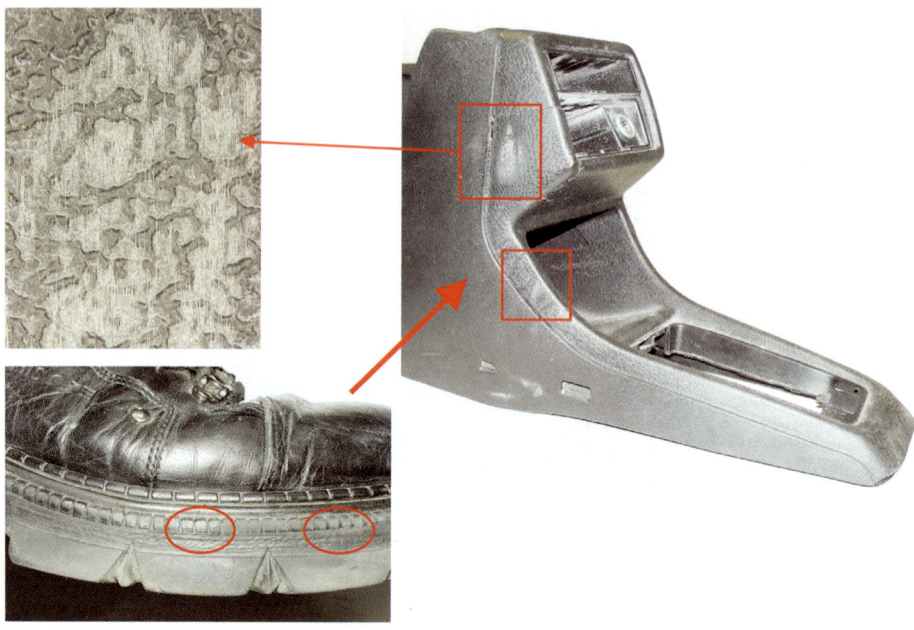

Figure 2.33
Plastic coating marks matching the material of the soles of the dead person's shoes on the left side of the middle console framed with red rectangles; the corresponding melting marks on the right side of the dead person's right shoe are framed with red ellipses (Case 4).

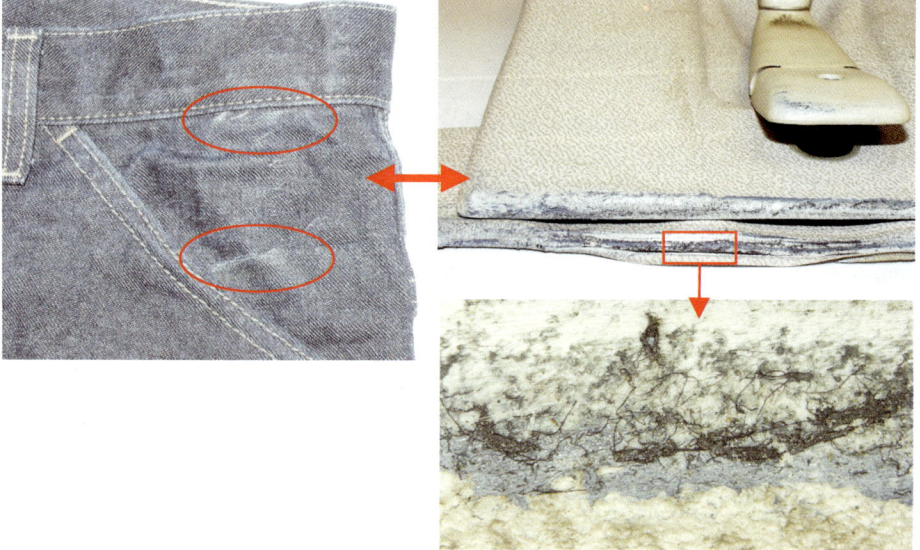

Figure 2.34
FPFs on the roof upholstery and the crank-operated sliding roof with embedded blue denim cotton fibers matching those of the front passenger's blue jeans; the corresponding plastic coating marks in the area of the left front pocket of this jeans are framed with red ellipses (Case 4).

dyed jeans (see Figure 2.34). The threads were running from the right to the left and vice versa, indicating how the person who caused this fusion mark moved when the car overturned in the course of the accident – but it must be pointed out that it could not be determined in which direction he moved first. Examination of the blue jeans revealed two plastic coating marks in the area of

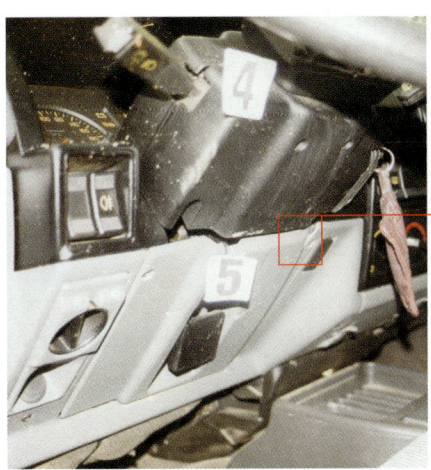
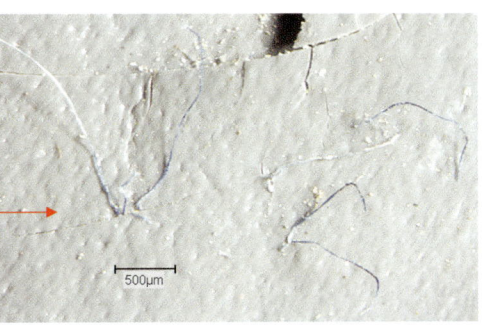

Figure 2.35
The lower parts of the dashboard; pinched blue denim fibers were found in the marked area just below the steering column cover (Case 5).

the left front trouser pocket, one running from the left to the right and the other vice versa. The transferred plastic materials matched the polyvinylchloride containing material of the roof upholstery and the sliding roof, respectively. The contact was, therefore, proven by a multiple cross-transfer of trace material.

CASE EXAMPLE 5

The driver of an imported car broke through a police check-point on a narrow road. When he tried to escape from the pursuing police cars, he made a risky passing maneuver on an un-gated grade crossing and immediately lost control of the vehicle. The car veered off the road and overturned in the bordering meadow on the left-hand side. The eyewitnesses saw two injured men crawling out of the car, but no one could tell the investigators who actually drove the car and, of course, both suspects gave no statement.

Fortunately, the crime scene officers had taken all plastic parts which were cracked or showed even the smallest change in texture in order to have the material examined in detail at the laboratory. No FPFs on the secured parts or plastic coating marks on the clothing of the suspects were found, although the car's body was severely damaged. But, as shown in Figure 2.35, several blue denim fibers were pinched in some cracks just below the steering column cover. This trace must have been caused by the driver's trousers.

The outcome of the examination was that the questioned driver must have worn indigo dyed blue jeans. Fortunately, only one of the suspects wore blue jeans, whereas the other one was dressed with a black one and the determination was easy to make. Most probably in a "normal" fiber case a result like "the person you are looking for must have worn a blue denim trouser" would have been shrugged off because such fabrics are so frequently encountered. But in the present case only two suspects were involved and the investigators just had to find out which of them drove the car; the "closed system" of the car made a quick and unequivocal assignment possible.

FPFs IN HIT-AND-RUN CASES

FPFs and related trace materials are also often formed on the outer parts of a car in collisions with pedestrians or bicycle riders, enabling the examiner to link the car to these persons. This shall be illustrated in the following case descriptions.

CASE EXAMPLE 6

It was already dark when a drunken pedestrian tried to cross a main road in the early evening hours on a day in January. In the middle of the four-lane road he was caught by a car and thrown back into the oncoming traffic, where he was run over by another car and killed. The driver of the car which hit the man first fled from the scene, but he went to the police together with his lawyer on the next morning. He told the police that he has not realized at the scene that he might have hit someone but he recognized the damages to his car when he came home. These statements were not very convincing and an examination of the car and the clothing of the victim was requested in order to determine if a link existed between them and to give an expert's opinion on the intensity of this contact.

An examination of the car revealed numerous damages: a broken fog-lamp and direction indicator on the left-hand side, a broken rear-view mirror at the driver's door and damages to the left A-pillar. According to these findings, it was evident that the car must have been involved in a crash and the driver must have been aware of it.

Two marks were present on the car which could be used to prove the contact with the pedestrian. A textile pattern, representing the typical twill lines of a denim fabric, was found just below the left headlamp. This could be assigned to the jeans of the victim (see Figure 2.36). On the A-pillar about 10 cm above the broken mirror, a plastic coating mark was present running from the front to the back. The transferred blue-green polyamide matched the upper material of the victim's jacket, a so-called "bomber jacket" with a blue-green polyamide fabric shell. Corresponding to the coating mark on the A-pillar, damage with fibers molten to some extent on the left elbow of the victim's jacket was noted. It was concluded that most probably the material on the car was melted off from that area (see Figure 2.37).

Figure 2.36
Textile pattern on a cover strip just below the left headlamp representing the twill lines of the victim's jeans (Case 6).

Figure 2.37
Plastic coating mark on the left A-pillar of the car and the corresponding damage with fibers molten to some extent in the area of the left elbow of the victim's jacket (Case 6).

While this was not a traditional "fiber" case, one must keep in mind that garments can produce a recognizable pattern on (dirty) surfaces (Kuppuswamy and Ponnuswamy, 1986; Ponnuswamy and Kuppuswamy, 2000) and some man-made fibers are themselves thermoplastic material and, therefore, can be transferred as a film as well.

Figure 2.38

Reconstruction of the crash with the bicycle (Case 7).

CASE EXAMPLE 7

A young female bicycle rider tried to turn from a main road into a side street on her left-hand side. Probably she did not recognize that an oncoming car – a BMW – was driving much too fast, however, the car ran into the side of the bicycle (see Figure 2.38). The rider was thrown over the hood into the windshield and collided with the female front seat passenger. The dead body of the bicycle rider was found approximately 60 m away from the place of collision. The driver fled the scene, but the car was found within some hours.

The examination of the car revealed a lot of trace material which proved the contact of the bicycle rider with the car. Plastic material from the bicycle was found on the hood and hairs and tissue were recovered from the windshield, but here the focus will be on the three FPFs: one on the front bumper just above the registration number plate, one on the BMW emblem (see Figure 2.39), and one on the right nozzle of the windscreen washer (see Figure 2.40). Colorless cotton and flax fibers matching the material of the victim's trousers were embedded in each of these FPFs. The fusion marks ran from the front to the back of the car, thus reflecting the movement of the victim.

Examination of the victim's trousers revealed a corresponding plastic coating mark, running from the top to the bottom, on the right side in the area of her hip (see Figure 2.40).

The transferred material matched that of the nozzle of the windshield washer which was made of a polyoxymethylene polymer (POM). Once again the contact was proven by a cross-transfer of trace material.

Figure 2.39
FPF with embedded colorless fibers on the BMW emblem (Case 7).

CONCLUSION

As shown in these examples, the analysis of FPFs and related trace material in accident reconstruction can be extremely successful. Indispensable prerequisites for such an outcome are a meticulous examination of the evidence and its preservation. The "closed system" of the accident provides a very tight context within which to interpret the results of the analysis. While this may not be the case in other crimes with trace evidence, the intellectual processes are similar.

86 TRACE EVIDENCE ANALYSIS

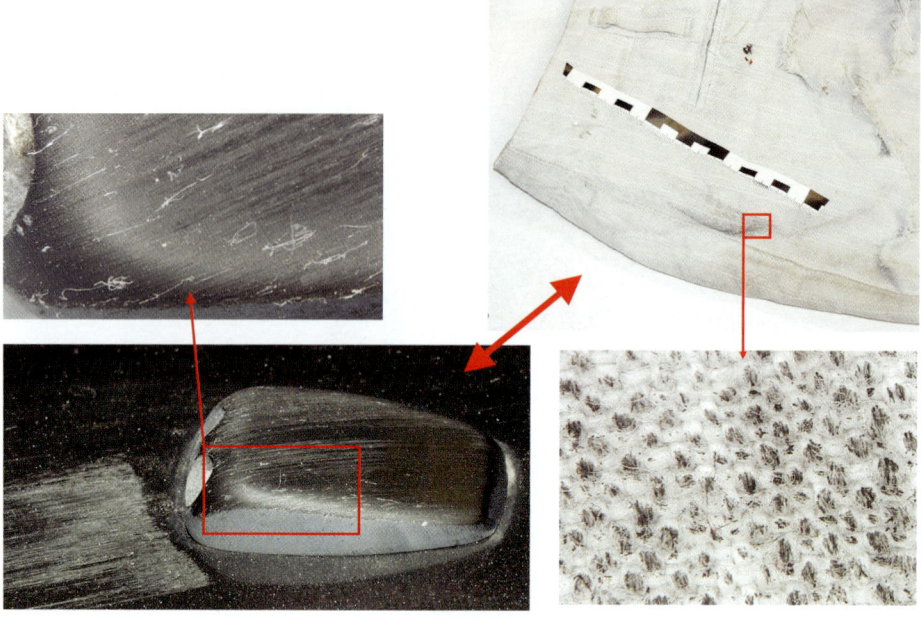

Figure 2.40
FPF on the right nozzle of the windscreen washer and the corresponding plastic coating mark on the victim's trousers (Case 7).

The orderliness of these reconstructions may provide a basis for the application of this kind of interpretation in other, less ordered, scenarios.

ACKNOWLEDGEMENTS

Special thanks are due to Dr. Herbert Pabst for valuable discussions and help and Bernd Maaz of the State Criminal Police Office of Saxony (Landeskriminalamt Sachsen) for co-working on the cases described above. The provision of some trace material and pictures presented in this chapter by Dr. Pabst is gratefully acknowledged.

REFERENCES

Bürger, H. (1977a) "Possibilities of determining the person steering a surface-craft or an aircraft at the time of accident (impact)," *Forensic Science*, 9, 5–12.

Bürger, H. (1977b) "Wer hat das Kraftfahrzeug zum Unfallzeitpunkt gelenkt? (Who drove the car at the time of the accident?)," *Der Verkehrsunfall*, 5, 99–101.

Bürger, H. (1989) "Grundlagen über die Entstehung von Abriebspuren an Bekleidung und KFZ-Innenteilen (Basic facts concerning the formation of FPFs and plastic coating marks)," *Verkehrsunfall und Fahrzeugtechnik*, 11, 301–306.

Jochem, G. and Pabst, H. (2000) "FPFs in traffic accident reconstruction," *Proceedings of the 8th Meeting of the European Fibers Group*, Cracow, Poland.

Jochem, G. (2001a) "Rekonstruktion der Insassen-Sitzverteilung in Unfallfahrzeugen (Reconstruction of the seating arrangement in cars involved in traffic accidents)," *Kriminalistik*, 5, 341–346.

Jochem, G. (2001b) "Traffic accident investigation – case examples and rarities," *Proceedings of the 9th Meeting of the European Fibers Group*, Helsinki, Finland.

Krauß, W. and Stritesky, K. (1993a) "Auswirkungen von Licht- und Wettereinflüssen auf textile Anschmelzspuren (Influence of sunlight and weather factors on fibers in plastic fusion marks)," *Archiv für Kriminologie*, 191, 99–106.

Krauß, W. and Stritesky, K. (1993b) "Rekonstruktion der Sitzordnung bei einem Motorradunfall anhand der Untersuchung textiler Anschmelzspuren (Reconstruction of the seating arrangement on a motorcycle by analysis of FPFs)," *Archiv für Kriminologie*, 192, 12–16.

Kuppuswamy, R. and Ponnuswamy, P.K. (2000) "Note on fabric marks in motor vehicle collisions," *Science & Justice*, 40, 45–47.

Lautenbach, L. and Schaidt, G. (1970) "Abdruckspuren an Schuhsohlen von Kraftfahrzeuglenkern (Impressions on the soles of the driver's shoes)," *Archiv für Kriminologie*, 146, 75–84.

Masakowski, S., Enz, B., Cothern, J.E. and Row, W.F. (1986) "FPFs in traffic accident reconstruction," *Journal of Forensic Sciences*, 31(3), 903–912.

Metter, D. (1978) "Die Rekonstruktion der Sitzordnung bei PKW-Unfällen (Reconstruction of the seating arrangment in cars involved in accidents)," *Archiv für Kriminologie* 162, 92–102.

Pabst, H. (1984a) "The textile-plastic fusing mark: guidepost to the car collision driver," *Proceedings of the 10th Triennial Meeting of the International Association of Forensic Sciences*, Oxford, UK.

Pabst, H. (1984b) "Microscopic differentiation of thermoplastics demonstrated by the microstructure of different black plastics in the interior of passenger automobiles," *Proceedings of the 10th Triennial Meeting of the International Association of Forensic Sciences*, Oxford, UK.

Pabst, H. (1992) "Anschmelzspuren (FPFs)," *Kriminalistik,* 8–9, 527–549.

Ponnuswamy, P.K. and Kuppuswamy, R. (1986) "Collision marks on plastic material on motor vehicles," *Journal of Forensic Sciences*, 31, 778–781.

Putnam, B. (2001) "Plastics in automobiles," in *Mute Witnesses – Trace Evidence Analysis*, ed. M.M. Houck. Academic Press, pp. 69–85.

Robertson, J. and Grieve, M. (eds.) (1999) *Forensic Examination of Fibers*, Taylor & Francis, London.

Schiller, W.-R. (1995) "Textilfasern in Anschmelzspuren (Textile fibers in plastic fusion marks)," *Kriminalistik*, 728–730.

Schwarz, W., Teige, K. and Brinkmann, B. (1985) "Die Rekonstruktion der Sitzposition durch Mikrospurenanalyse (Reconstruction of the seating position by means of microtrace analysis)," *Z. Rechtsmedizin*, 94, 213–218.

Turner, D. (1999) "PC-Crash – A new tool for vehicle accident investigation," *CONTACT*, 27, 4–6.

CHAPTER 3

AN INTELLIGENCE LED INVESTIGATION USING TRACE EVIDENCE

Ray Palmer
The Forensic Science Service, Huntingdon, UK

INTRODUCTION

The East Coast of England can be a particularly bleak, cold, and uninviting location in the month of February. It was here on a beach that a grim discovery was made that started a murder investigation, which exemplified the value of trace evidence both in the investigative and corroborative phases of such an enquiry.

Early in the morning of February 3rd, a man walking his dog noticed that a large object had been washed up just beyond the tidal line of the beach. On closer inspection, it became evident that this "object" was the body of a naked male around which chains had been wrapped and these in turn threaded through green metallic painted barbell weights. The chains themselves were padlocked. The dog walker immediately contacted the police and the subsequent murder investigation began.

The scene was cordoned off and the area around the body was meticulously searched; initially, nothing of significance was found. At this stage, it was unclear whether the body had been dumped offshore at this location or at another deposition site. The body was removed from the scene and taken to the mortuary for a postmortem examination.

The postmortem revealed that the deceased had not drowned, but had died from blunt force injury to the head. There were no apparent external injuries, which would have been likely to bleed and thus leave biological evidence. It was estimated that the body had been in the sea for approximately 4 days. The inside of the mouth and upper throat appeared to have been packed with some form of foreign tissue – a finding which was later to prove significant for several reasons (see Figure 3.1). This gag was removed and submitted to the laboratory for further investigation, as were the chains, weights, and padlocks.

By now, the investigation had expanded beyond the initial scene, and a single barbell weight of similar color was found on a beach cove approximately 1 mile from the site where the deceased was found. The pattern and speed of tidal

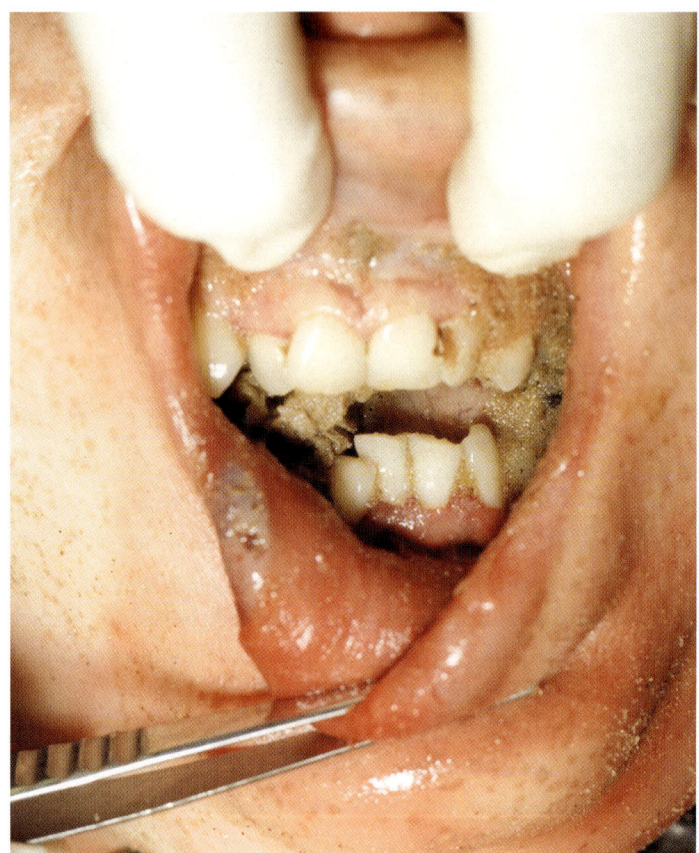

Figure 3.1
Photograph showing debris inside the deceased's mouth.

streams in this area were consistent with the deceased having been deposited in the water near the cove and then having been washed up at the site where he was found (see Figure 3.2). The recovered weight was submitted to the laboratory for comparison with those chained to the body of the deceased (see Figure 3.3).

In this type of case, the first priority for the police investigative team is to establish the identity of the deceased; only once this is achieved, can any murder enquiry progress. In this particular case, establishing the identity of the deceased proved easier than anticipated as a "hit" with his fingerprints was made on the national database. It was established that the deceased had been convicted of a minor offense some years previously and his fingerprints had been taken at the time of his arrest.

Subsequent enquiries revealed that up until the discovery of his body, the deceased shared a home with his ex-wife. Investigators learned that their relationship was somewhat tempestuous and, according to his ex-wife, the victim had a habit of "disappearing" after arguments. The deceased had done so following what appeared to be the most recent of a series of frequent arguments. The

Figure 3.2
Photograph showing location of victim's body (marked by "X") in relation to original deposition site (marked by arrow).

Figure 3.3
Barbell weights attached to the deceased.

ex-wife told the investigating officers that the victim had left over a week before the discovery of his body, but that length of time was not an unusual occurrence. The task for the investigating team was now to try and establish the last movements of the victim prior to his death and subsequent discovery on the beach.

In such circumstances, the investigative team will call upon the forensic laboratory to help assist in determining the movements of a victim by providing any information from the analysis of evidence which may provide intelligence. "Intelligence" in this context refers to information, which may steer the investigation in a given direction, such as identifying likely suspects or providing indications of the environment(s) in which the victim may have been prior to their discovery. In this particular case, it was initially thought that since the victim was found naked and had been immersed in water for some time, the prospect of obtaining any forensic intelligence would be slim. The gag, weights, chains, and padlocks obtained from the deceased, therefore became the priority for the laboratory examination.

INITIAL LABORATORY EXAMINATION

The first task at the laboratory was to determine the nature of the gag recovered from the victim's mouth and throat. This appeared to be a hopeless task because, due to its immersion in water and having been in the victim's mouth, the gag presented itself as a white amorphous mush. As the gag dried, it was gently teased apart to determine its structure and identity. The gag had been folded up on itself in what appeared to be several wad-like structures. It was hoped that the folds (see Figure 3.4), as well as the fact it had been lodged in the victim's mouth, would have promoted retention of any debris present on it prior to its use, even under the extreme conditions of immersion in the sea. This proved correct and ultimately the following were found within the folds of the gag:

- three turquoise non-delustered acrylic fibers,
- three dark blue crenulated viscose fibers,
- three pale yellow acrylic fibers,
- one blue acrylic fiber,
- one brown/green acrylic fiber, and
- three cat hairs.

The now-dry gag itself was of equal interest: it was carefully unfolded and revealed itself to consist of at least two sheets of white, non-embossed, two-ply paper toweling with clearly visible perforations.

This information was of obvious interest to the police: assuming the gag had been placed in the victim's mouth by his killer or accomplice, these findings potentially provided information about the perpetrator's and victim's overlapping environment(s).

The chains, padlocks, and weights wrapped around the deceased's body were also examined for any trace evidence; however, none was found. The barbell weights themselves were painted a green metallic color. Comparison of the weight found at the deposition site showed it to be of the same design, make, and construction of those chained to the deceased's body. Comparative analysis of the paint on the barbells from both sites showed them to be indistinguishable. Although seemingly innocuous at this point in the investigation, these findings were to prove crucial later on.

Figure 3.4
Mouth debris dried and teased apart, revealing paper toweling. Circles show position of perforations. Arrow indicates 2-ply construction.

The police had begun to interview known associates of the deceased and it was decided to seize any textile material resembling the description of the color and type of fibers recovered from the gag from these individual's homes. As a consequence, many items were submitted to the laboratory, but subsequently eliminated, as potential sources for the fibers in question. In addition, a sample of any paper toweling found at the home of these associates was taken for further comparison with the gag. Again, none were a "match" to the gag.

POLICE INVESTIGATION

After tracing and interviewing the known associates of the deceased with no further investigative leads, the police turned their attention to the deceased's ex-wife and her brother. From interviews it had become obvious that there was no love lost between the deceased and his former brother-in-law, who had made it clear that he was far from happy about the relationship and domestic arrangements existing between the deceased and his sister. Both the ex-wife and her brother maintained they were unable to speculate on the deceased's whereabouts prior to his discovery. The beach where the deceased's body was found was over 100 miles from their home and, moreover, in a county in which the brother-in-law claimed he had never been. It was this claim, which cast doubt over his innocence, as a routine check of the brother-in-law's mobile phone records showed that his phone had transmitted a call from the very area in which he claimed he had never visited. Furthermore, this call had been transmitted within the time-frame of the deceased's disappearance. As a consequence, the investigation began to focus more closely not only on the brother-in-law, but also on the ex-wife.

The decision was taken to examine the houses of both of these suspects to determine whether there was any evidence of the deceased being assaulted there and/or whether there were any sources for the fibers found in the gag of the deceased or for the gag itself.

Due to the remote location of the beach where the deceased was discovered from where the two suspects lived, if the deceased had been killed in or near one of the homes of the suspects, then he would have had to have been transported the considerable distance to be dumped at sea. A vehicle of some type must have been used to transport the body. The deceased's ex-wife did not drive; the brother-in-law, however, did own a car and maintained the deceased had never been in it. The brother-in-law's car was seized and submitted to the laboratory for further examination.

HOUSE EXAMINATION

Because of the relationships between the two suspects and the deceased, two forensic search teams, one for ex-wife's and one for the brother-in-law's house, were deployed. This would help in refuting any future claims of contamination between the suspected sources. The house where the deceased and his ex-wife had lived during their marriage yielded the most interesting findings.

On entry to this house, it was evident that it was also home to several cats; cat hairs had been found on the gag. While at face value this finding in itself did not appear to be significant, it nevertheless gave some indication that more potential evidence may be present within. The deceased's ex-wife was a prolific knitter and many balls of wool were present throughout the house, as was a knitting machine, and many items of knitted clothing were strewn throughout the various rooms. In order to facilitate the task of identifying any potential sources for the fibers in the gag, any items of clothing or wool yarn which could not be eliminated visually were sampled on site and the constituent fibers examined microscopically. In the end, two knitted sweaters were identified for submission

Figure 3.5
*Cat hairs at 200×
magnification.*

to the laboratory for further examination, as were samples of fur from all of the cats (see Figure 3.5).

The home of the brother-in-law was similarly examined but no potential sources for any of the fibers present in the gag were found. No blood staining indicative of any violent assault was found in either of the houses; because there was no evidence of bleeding-type injuries to the deceased, this was perhaps of limited significance.

CAR EXAMINATION

The brother-in-law's car was received at the laboratory and the meticulous, time-consuming task of its examination began. A strategy for the search of this vehicle was established:

- examine for any body fluids attributable to the deceased,
- search for fibers with common origin of those found in the gag,
- search for paint attributable to that on the barbell weights attached to the deceased, and

- presence of any sand/soil linking the vehicle to the area where the deceased was discovered and/or the deposition site.

An initial visual examination for any body fluid staining proved negative. In order to effectively search for target trace material, surface debris had first to be recovered from the upholstered surfaces of the passenger compartment and the interior of the trunk before it could be searched. This was achieved by a combination of vacuuming and taping.

Taping involves taking a length of clear adhesive tape (such as cellophane tape), pressing this adhesive-side down onto the surface in question, and applying pressure along the length of the tape. The tape is then lifted off and reapplied to the next area of the surface in question. This procedure lifts off debris, such as fibers, hairs, and other small particulates from the recipient surface. The tape is then fixed onto a clear acetate sheet, effectively sealing in the recovered debris, and preserving any potential evidence. Since the tape and the acetate sheet are transparent, the tape can then be searched using a low power microscope. Any trace evidence of interest which is identified, can be removed by cutting a "window" or flap in the tape around the object of interest and then removing the particulate in question using fine forceps. The item is then mounted in a medium appropriate for its further examination and comparison. When using this technique of trace evidence recovery, it is important not to "overload" the tape with debris. If the tape is overloaded, the adhesive properties of the tape are reduced and the recovery efficiency is compromised. In addition, such overloaded tapes are extremely difficult to search. For this reason, numerous tapings are often used to recover trace evidence from a comparatively small area.

Vacuuming, as the name suggests, involves the use of what is essentially a vacuum cleaner modified with a filter unit inserted in the vacuum flow. As debris is sucked from the surface in question, it is collected in the filter assembly. The filter is removed and the debris placed into a clear Petri dish or similar receptacle for subsequent searching with lower power microscopy. This method of debris recovery is very useful when there are large particulates present in the debris (such as soil or sand), which would make the use of taping difficult and inefficient. The disadvantages of vacuuming are that it is indiscriminate and can (ironically) be too efficient: It can collect too much debris, often including items that are not relevant to the time-frame of the incident, making a microscopical search of the debris difficult. In addition, it is vitally important that the collection surfaces of the vacuum are scrupulously cleaned after each use.

The floors of the suspect's vehicle were first vacuumed and the debris from each placed into separately labeled containers. The carpeted regions of each floor were then taped, as were the upholstery of the seats, seat belts, and the interior surfaces of the trunk. Each of the tapings was labeled with the location in the vehicle where the surface debris was removed.

The under-carriage of the vehicle was also examined and it was noted that large clumps of sand were present in the wheel well recesses and on the inside hollow of the front bumper and sills. Again, this debris was placed into appropriately labeled receptacles. Once the examination of the vehicle was finished, there were literally dozens of tapings and numerous containers of debris to be searched.

RESULTS OF LABORATORY EXAMINATION
ITEMS RELATING TO THE HOUSE

The two sweaters seized from the house were submitted to the laboratory and reference samples (so-called "control samples") were collected by removing constituent fibers from the garment and directly mounted onto a microscope slide. These fibers were then compared microscopically with the fibers recovered from the gag (see Figure 3.6) It is important in any scheme of comparison that the most rapidly-discriminating tests are carried out first so that any exclusions can rapidly be made. This sorting approach makes more efficient use of the more-discriminating but time-consuming instrumental techniques. In fiber comparison and analysis, the first technique employed is that of comparison microscopy. This is as, the name suggests, two microscopes connected by an optical bridge allowing direct comparison in the same field of view of two separate samples. The color and microscopic features of the fibers can be directly compared side-by-side, as can the optical properties of the fiber, such as fluorescence colors, under different wavelengths of light. Additionally, the generic class of the fibers (e.g. acrylic, nylon, etc.) can be determined at this stage of the analysis. In the event that no differences are detected between the control sample and the suspect fibers, then further instrumental comparison is employed.

The first of these techniques is known as microspectrophotometry (MSP). Although the human eye is capable of extraordinary color differentiation, it is subjective and, in some instances, can be "fooled" into believing that two different colors are the same when viewed under certain lighting conditions. The technique of MSP provides objective color discrimination that cannot be provided by the eye alone. It achieves this by examining the absorption characteristics in the visible spectrum, which give the dyed fiber its particular color. Another technique called thin layer chromatography (TLC) can be used to separate out the dye components from the control and suspect dyes for direct comparison. In addition to the color comparison, Fourier transform infrared spectroscopy (FTIR) can be used to directly determine the chemical composition of fibers.

Figure 3.6
Searching the debris for fibers using low power microscopy.

If, after all these tests have been carried out, no differences are found between a suspect and control fiber sample, the fibers are said to be indistinguishable. In such a circumstance, it is, therefore possible that the suspect fiber(s) originated from the same garment as the control sample.

Such a scheme of analyses was performed on the fibers recovered from the mouth debris and control fiber samples from the submitted sweaters. The single blue acrylic fiber recovered from the mouth debris was found to be indistinguishable from the fibers comprising one of the sweaters, and the single brown/green acrylic fibers also from the mouth debris, was found to be indistinguishable from the fibers comprising the other sweater.

The cat hairs recovered from the mouth debris were microscopically compared to the samples of hair obtained from the cats at the house. None of the cats could be excluded as a source of these hairs.

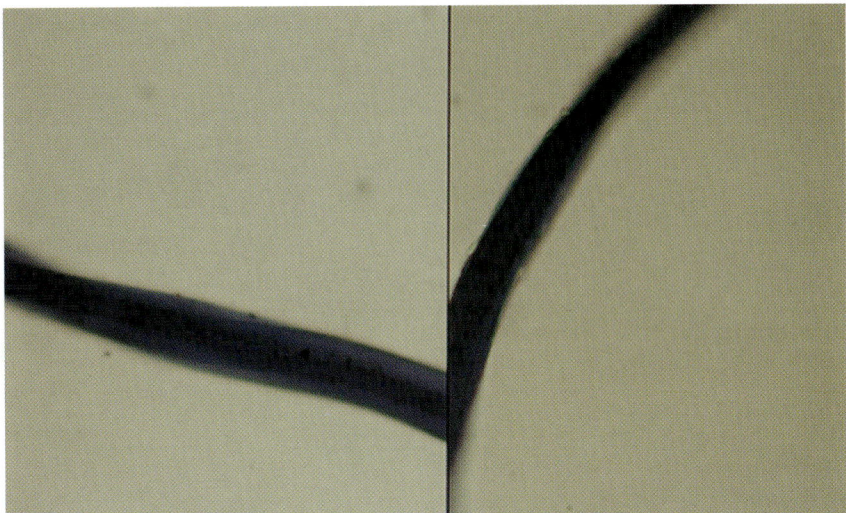

Figure 3.7
White light comparison of fibres from the car trunk (right) and mouth packing (left).

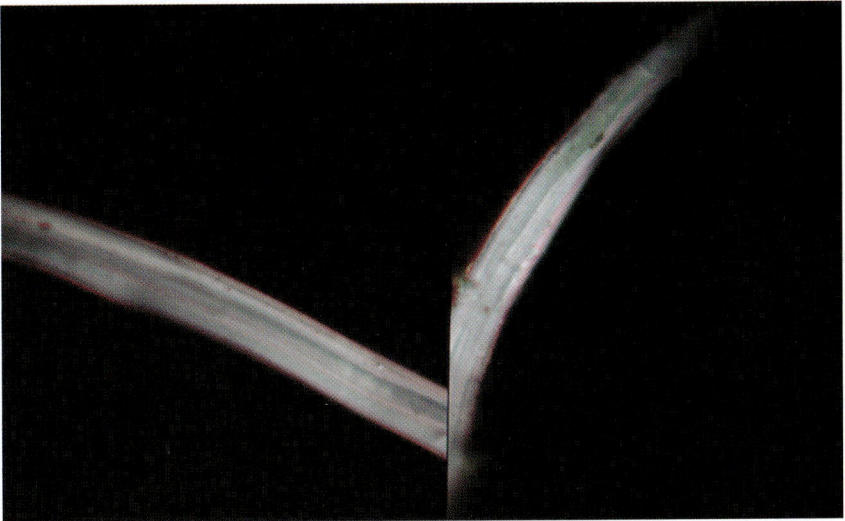

Figure 3.8
Comparison under polarized light between fibres from car trunk (right) and mouth packing (left).

DEBRIS FROM THE CAR

After a painstaking search of the debris from the car, three dark blue viscose fibers were found in the debris of the trunk; these were found to be indistinguishable, using the previously described methods, from the three dark blue viscose fibers recovered from the mouth debris of the deceased (see Figures 3.7 and 3.8). The significance of this finding was somewhat problematical, as no common source for these fibers had been identified at this stage of the investigation.

Figure 3.9

Comparison between paint flake recovered from car (left) with sample from barbell weights attached to victim (right). Note: these fragments are less than 1 mm in length.

More significantly, a single tiny fragment of green metallic paint was found on the rear passenger floor. This fragment was an acrylic type of paint which was indistinguishable in terms of its color, microscopic appearance, and chemical composition from the paint on the barbell weights strapped to the deceased's body with chains and from the weight found at the deposition site (see Figure 3.9). In addition, analysis of the sand from the car had revealed it to be similar to that on the coastline where the deceased was discovered, but completely dissimilar to that of the coastline which the brother-in-law claimed the sand originated.

CHAINS/PADLOCKS

The chains and padlocks found on the deceased were examined and the construction features were noted and, as luck would have it, turned out to be unusual. The results of this examination were passed to the police team who contacted the manufacturers. The results of this enquiry revealed that the chains and padlocks were sold by a particular chain of stores, one of which was literally yards from the brother-in-laws' address. Importantly, the store records showed that the last time such items had been purchased had been during the time-frame of the deceased's disappearance. Unfortunately for the police, the purchase had been made with cash and the security cameras had been inoperative during the purchase. Nevertheless, it was felt that such evidence, even though circumstantial, was beginning to corroborate the physical evidence in the case.

By now, the analysis of the trace evidence in the case was beginning to present quite specific links between the deceased, the car of the brother-in-law, and his house.

SIGNIFICANCE OF THE FINDINGS

In such cases where physical evidence is found which ostensibly provides evidence of association between individuals, or individuals and places, it is important that some form of assessment of the strength of such an association is made.

In the case of the fiber evidence it was important to attempt to assess:

- the single fiber matches found between items of clothing from the deceased's home address,
- the cat hairs, and
- the significance of fibers from the boot of the suspect car matching those found in the mouth debris of the deceased – with no demonstrable source.

A simple and not unreasonable defense relating to the fiber findings in a case such as this would be that any matching fibers apparently demonstrating associations between an accused and a victim are purely coincidental. Such an argument will maintain that since textile materials are mass-produced, then it would not be unreasonable to suggest that any fiber matches found are the result of coincidental contacts with garments of identical fiber construction, but unrelated to the person(s) in question. The question regarding the evidential significance is: How likely is it that one would encounter a given fiber color/type combination on a random surface?

In order to help evaluate these findings, one can turn to specific empirical studies that have been carried to help address such interpretative issues, such as those as pertinent to this case. The first of these types of study are called "target" fiber studies. These involve looking for a given fiber of a particular color in debris collected from surfaces, which are subjected to random contacts from the population at large. Examples of such studies are those which have looked for "target" fibers on car seats (Jackson and Cook, 1986), movie theatre seats (Palmer and Chinherende, 1996), and seats in bar rooms (Kelly and Griffin, 1998).

The results of these studies show that where the fiber type and combination is not ubiquitous (such as denim cotton fibers or certain types of black cotton fibers (see Grieve, 2001; Roux and Margot, 1997)), it is unlikely that a given fiber will be found on a random surface by chance. Where more than one fiber color/type combination is searched for (e.g. from a garment composed of multiple fiber types), the data shows coincidental matches to be extremely remote.

The second type of study that is also helpful in evaluating the significance of fiber examination findings are fiber population studies. While target fiber studies provide information on how likely it is to encounter a particular fiber type/color combination, the opposite question can also arise: How common is a particular fiber? While to a certain extent this question can be addressed by the experience of an analyst, fiber population studies seek to provide some empirical data to assist in answering this question.

These studies quite simply examine and catalog the relative proportions of fiber type/color combinations present on surfaces exposed to random contacts in the population. Examples of the types of surfaces examined are various outdoor surfaces (Grieve and Biermann, 1997), car seats (Roux and Margot, 1997), and T-shirts (Massonet *et al.*, 1998).

The results of these and other studies have shown that the commonest fiber types encountered are blue and black cottons but that manufactured fibers of a given color were comparatively rare. The information obtained from these studies tends to confirm the experience of analysts that the finding of a given target fiber, even those not particularly uncommon, is a comparatively rare event.

In attempting to assess the significance of the fiber matches relating to the various locations in this case, we can consider alternative explanations for the findings.

The presence of two single fibers in the mouth debris both matching a separate garment in the brother-in-law's house, as well as cat hairs in the debris which cannot be excluded as having come from the cats in the house, could be explained by:

1. the paper toweling constituting the mouth debris had been within the environment of the house, or
2. the mouth debris had been in an environment in which items of identical fiber composition and similarly colored cats were present.

Again, we can ask ourselves what are the explanations for finding a number of given fiber type/color combinations in the mouth debris from the deceased matching a similar number present in the trunk of the suspect vehicle without any demonstrable source for these fibers. It would appear there are two possible explanations:

1. a defined source of these fibers had been in contact with both the trunk of the suspect vehicle and the paper toweling constituting the mouth debris from the deceased, or
2. two separate and unrelated items of identical fiber composition had each been in contact with the mouth debris and the suspect car boot.

From our knowledge of the target fiber and fiber population studies, as well as subjective assessments of the chances of finding a given combination of items, it is the author's opinion that the first explanation for the findings in each case is more likely than the second. Given that the deceased had last been seen wearing a dark blue suit, and that information from a database suggested that the fiber type/color of the fibers from the trunk and mouth debris were commonly used in (but not exclusively) jackets and trousers, then it can be seen that these findings provide further evidence in this case.

The findings relating to the paint evidence in this case proved less problematic, in that demonstrable sources for the paint fragments found in the vehicle were found. The paint evidence on its own therefore showed clear associations between the suspect vehicle, the deposition site and the deceased.

SUMMARY OF THE EVIDENCE

The nature and extent of the physical evidence in this case provided several links to the suspects, deposition site, and the deceased. While it can certainly be argued that none of the physical evidence found was, on its own, conclusive, it would be difficult to explain the summation of these findings as series of coincidences. When the trace evidence in this case was combined with the evidence produced by the investigative efforts of the police, a compelling case against the ex-wife and particularly the brother-in-law was constructed.

The suspects confessed to the crime before the trial. They admitted that following an argument at the deceased's home, a fight occurred between the deceased and the brother-in-law. This resulted in the deceased receiving a blow to the head, killing him. On realizing that he was dead, the suspects removed his clothing, packed his orifices with the remains of a paper towel ("to prevent any leaking") and moved the body to an outhouse. The body was later placed in the trunk of the brother-in-law's car, along with the decedent's clothing, and driven to the deposition site. There, the deceased was wrapped in the recently purchased chains, which were threaded through the barbell weights; the weights had come from the suspect's garage. The body was then placed on an inflatable mattress, dragged along the beach, and was pushed out to sea.

The clothing of the deceased was buried nearby and never recovered.

SUMMARY

The circumstances of this case, and the subsequent laboratory and police investigation exemplified how the analysis of trace material can have a profound influence on the nature and outcome of a major enquiry, from the initial intelligence phases through to the corroborative phases, by a teamwork based approach.

In addition, it demonstrated the quandaries which the trace analyst can find him/herself in when trying to assess the significance of their findings and the data from studies which can help in the interpretation of their findings.

REFERENCES

Bruschweiler and Grieve (1997) "A study on the random distribution of a red acrylic target fibre," *Science & Justice*, 37, 85–89.

Grieve and Dunlop (1992) "A practical aspect of the Bayesian interpretation of fibre evidence," *Journal of the Forensic Science Society*, 32, 169–175.

Grieve, M.C. (2001) "The evidential value of black cotton fibres," *Science & Justice*, 41, 245–260.

Jackson and Cook (1986) "The significance of fibres found on car seats," *Forensic Science International*, 32, 275–281.

Kelly and Griffin (1998) "A target fibre study on seats in public houses," *Science & Justice*, 37, 39–44.

Palmer and Chinherende (1996) "A target fibres study using cinema and car seats as recipient items," *Journal of The Forensic Science*, 41, 802–803.

Roux, C. and Margot, P. (1997) "The population of textile fibers on car seats," *Science & Justice*, 37, 25–30.

CHAPTER 4

THE VALUE OF SOIL EVIDENCE

Thomas J. Hopen
Bureau of Alcohol, Tobacco, Firearms and Explosives
Forensic Science Laboratory – Atlanta
Arson and Explosives Section
Atlanta, GA, USA

INTRODUCTION

Unfortunately, soil evidence is not considered to have much, if any, evidentiary value by many forensic laboratories throughout the US. They usually place soil in a broad generic class and say "Well, most of the soil in our state is geologically the same and, therefore, we cannot do much with it". Or, some might go a step further and say "There are only two (or three, etc.) geological types of soil in our state and we cannot tell you much more than from what region the soil may have come". Nothing could be further from the truth than the assumption that soil has an ubiquitous consistency within a given state or geologic region and has no associative evidentiary value.

Although soil may have the same geological provenance within a region it may vary significantly in its makeup over short distances, both vertically and horizontally. These changes in the soil makeup can be attributed to a number of factors. Some of these factors may include:

- The natural weathering of the soil due to its location and the environmental conditions the soil is exposed to over time.
- The physical alteration of the soil due to tilling, grading, and/or digging which exposes and mixes in "fresh" soil.
- The mixing of two different soils together to give the soil a different consistency or other physical properties.
- Adding natural amendments to a soil such as one might do to a flower bed and/or lawn with lime (limestone), sand (quartz-based or limestone-based), gypsum, sulfur, vermiculite, peat, pumice/perlite, manure and/or humus.
- Indigenous additions to a soil such as pollen grains, phytoliths, plant fragments, etc.
- Addition of man-made contaminants such as paint chips, paint spheres, glass fragments, brick fragments, cement particles, shingle stones, welding spheres, and other types of debris which can be found in soils around homes, businesses,

and structure/construction sites. Also, one must remember that soil along roadways usually contains rubber particles, reflective glass spheres from traffic signs/paint, marking paint, automobile paint chips, and pavement particles (asphalt and/or cement particles).

While with the Alabama Department of Forensic Sciences the author worked a number of cases in which soil evidence played a significant part. Three cases are briefly described below and the fourth case is provided in more detail to illustrate the value of soil evidence.

CASE 1

On a warm summer evening, a burglary was attempted on a business with the perpetrator trying to gain entrance through a foundation crawlspace having a dirt floor. A man meeting the description of the suspect was stopped a block from the business by police and taken in for questioning. The suspect wore no shirt and his jeans were soiled. When questioned as to how his jeans became soiled, he said he had been playing baseball at a local park several blocks away. His clothes were submitted to the laboratory along with soil samples from underneath and around the perimeter of the business as well as from the park. Examination of the jeans revealed several clumps of soil deposited on the inside of the front waistband. Analysis and comparison of the soil from the jeans with the reference samples revealed it to be consistent only with the soil from underneath the business. The suspect pled guilty when confronted with the evidence. How did the soil get inside the waistband? It is believed that when the suspect was crawling on his belly, and with no shirt to block it, the waistband of the jeans scooped the soil up. His perspiration moistened the soil and helped tack it to the inside material of the jeans.

CASE 2

Clothing from a rape suspect was submitted to the laboratory along with soil samples from the rural field where the incident was alleged to have taken place. Examination revealed the knee areas of the jeans to be soiled with dark brown material. The material was very hard and immediately broke apart when placed in water. The aqueous dispersion was filtered though a fine sieve and a few mineral grains were recovered along with numerous trilobal acetate fibers. The suspect told the police that the jeans had been soiled when he was working in the flower beds in front of his house.

Samples submitted from the flower beds revealed them to be dissimilar to the soil material on the jeans. Further investigation revealed a local farm was spreading its manure on the field and the feed being used at the farm contained acetate fibers as filler. The case went to trial and the suspect was found guilty.

CASE 3

A trucking company located in a low-lying swamp area was burglarized and clothing articles from several suspects were submitted to the laboratory. Soil recovered from the shoes of the suspects proved it to be consistent with soil from the trucking company. The soil from the trucking company was rich in Foraminifer and was not consistent with the low-lying, swampy soil from the surrounding area. Further investigation revealed the soil from the trucking company to have been imported from a neighboring county to raise the soil level and support the trucks. The suspects pled guilty.

CASE 4

This case is being provided in more detail and will demonstrate how soil evidence can assist an on-going investigation.

THE CRIME

Late one evening in the spring of 1983, a young woman named Alice Redmond was reported missing by her husband. Co-workers told investigators that the woman was last seen leaving at the end of the day with Mark Miller, another co-worker. Alice worked as a court reporter at the courthouse and Mark worked in the law library at the courthouse. Mark was actually on probation at the time for burglary and theft and was working at the courthouse so he could pay off court-ordered restitution to the victims.

Upon returning to work the next morning, Mark was questioned by the police and stated that he and Ms. Redmond had just rode around and talked the previous evening. However, Mark's sister, who lived approximately 50 miles away near the Alabama/Georgia state line, was also questioned and informed the authorities that her brother had stopped by her house late that night. She said that Mark told her he had taken Ms. Redmond across the state line, stabbed her, and had buried the body where no one would ever find her. He was taken into custody and charged with kidnapping and murder.

THE ANALYSIS

The day after the victim was reported missing, her car was found in the parking lot of a local hospital (see Figure 4.1). The car was impounded and examination of the vehicle at the police department revealed numerous soil deposits in the wheel well areas of the car and on the under-carriage. The majority of the soil was a brownish-colored mix of mineral matter and plant fragments (see Figure 4.2). Several minute clumps of reddish soil were found on top of

Figure 4.1

Victim's car found at hospital parking lot. Note the brownish-colored soil in and around the wheel well.

Figure 4.2

Brownish-colored soil collected from the victim's car.

the brownish-colored soil, with the largest clump of reddish soil being approximately 1 cm in diameter.

Other items of evidence were submitted for examination, including a pair of blue jeans that the suspect was believed to be wearing the night of the incident. There were bloodstains noted on the jeans which serological analysis revealed to be the same blood type as the suspect's. An abundant amount of pollen grains, mainly pine pollen, were found adhering to the lower portion of both pant legs. During springtime in the southeastern US, the presence of pollen grains, especially pine pollen, is not uncommon. However, the abundance of pollen grains found on the clothing suggested that the clothing had recently been in the vicinity of a rural area. Examination of the other items of clothing failed to reveal anything of apparent evidentiary value.

Routine questioning of Alice's husband revealed that the couple had attended a motorcycle race on Sunday. At the race they had parked the car in a muddy

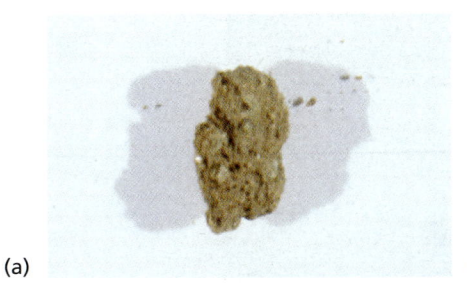 

Figure 4.3
Reddish-colored soil collected from the victim's car (a) and a reference sample submitted from Crowhop, Georgia (b).

field and got stuck. Samples were collected from the field and compared to the soil samples collected from the car. This comparison revealed that the brownish-colored soil had a similar composition to the samples collected from the field.

Investigators were disappointed since they had hoped the large deposits of brownish soil would help them to indicate where the body might be found. The soil did provide a key piece of information. The brownish-colored soil was deposited on Sunday evening and Alice drove the car to work on Monday. She reportedly left Monday evening with Mark Miller and the car was recovered on Tuesday morning. Therefore, the reddish-colored soil discovered atop the brownish soil was deposited sometime after Sunday evening and before Tuesday morning. Everyone now wondered if this minute amount of reddish soil might be a key to where the body was buried.

There are a number of analytical schemes for conducting soil examinations and comparisons. However, the first step in any soil examination scheme is to note the color and texture of the dry soil samples utilizing a gross visual examination and detailed stereomicroscopical examination. Any artifacts or contaminants, such as glass, paint, etc., that are noted would usually be isolated at this time. The investigators in this case submitted a number of soil samples for comparison purposes. Also, in the normal course of duties, the author had previously collected a number of reference soil samples from around the state. Comparison of the soil samples revealed none of the reference samples within the state to be similar to the minute reddish-colored soil from the victim's car. However, one of the soil samples collected by the investigators from a dirt road off Georgia Highway 103 across the state line near Crowhop, Georgia, had striking similarity and thus could not be eliminated as a possible source of the reddish soil on the victim's vehicle (see Figure 4.3). A map showing the area of Alabama along I-85 to the Georgia border and a corresponding soil profile map are provided in Figures 4.4 and 4.5. A map showing the Crowhop, Georgia area is provided in Figure 4.6.

The soil samples were next wet sieved into several different mineral grain size fractions. This process involves placing the soil in a stacked series of sieves, each one having a finer mesh than the one above it; water is used to move the soil

Figure 4.4
Map showing Montgomery, Alabama and the area of the state along I-85 towards the Georgia border.

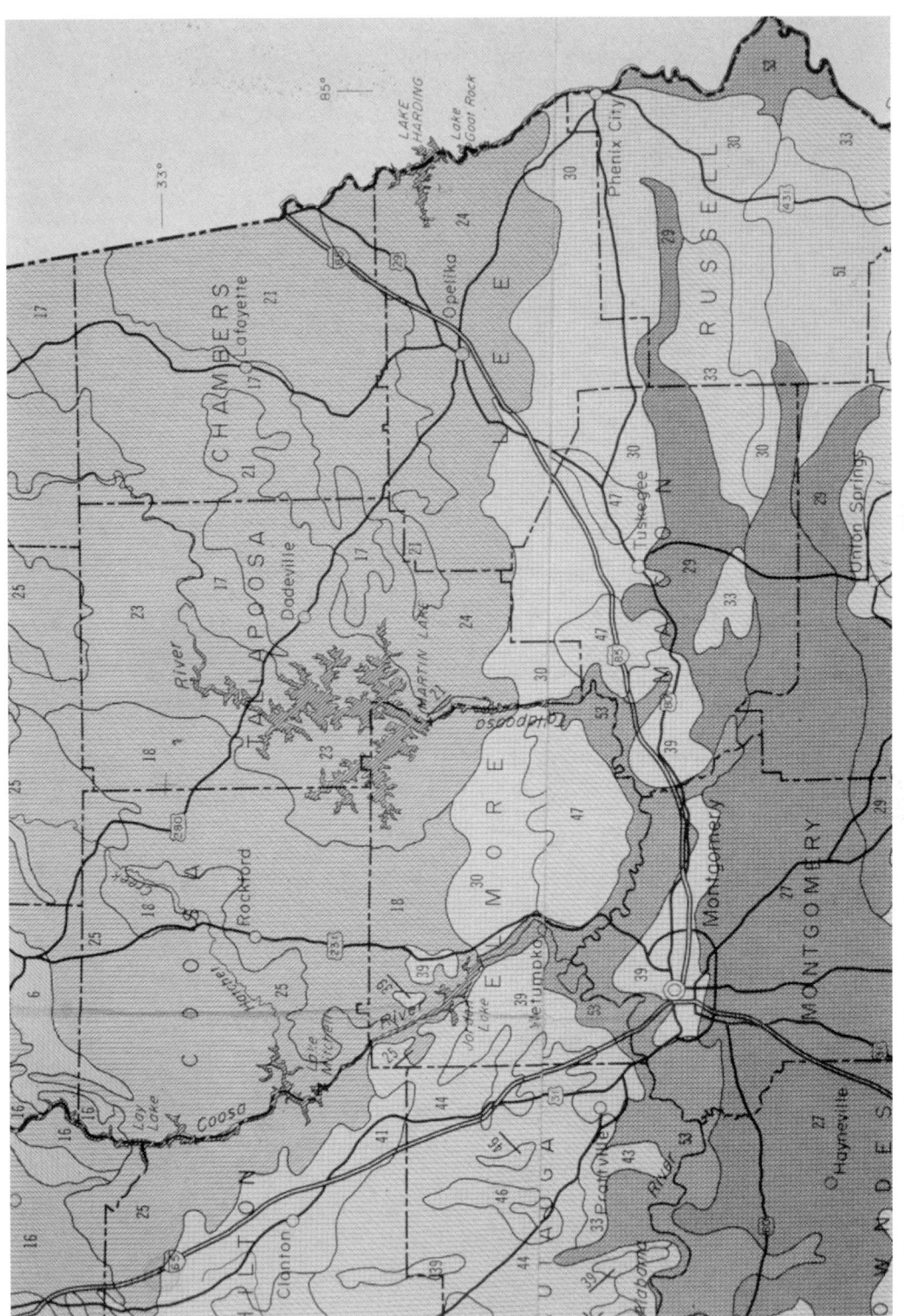

Figure 4.5
Soil profile map of the area shown in Figure 4.4.

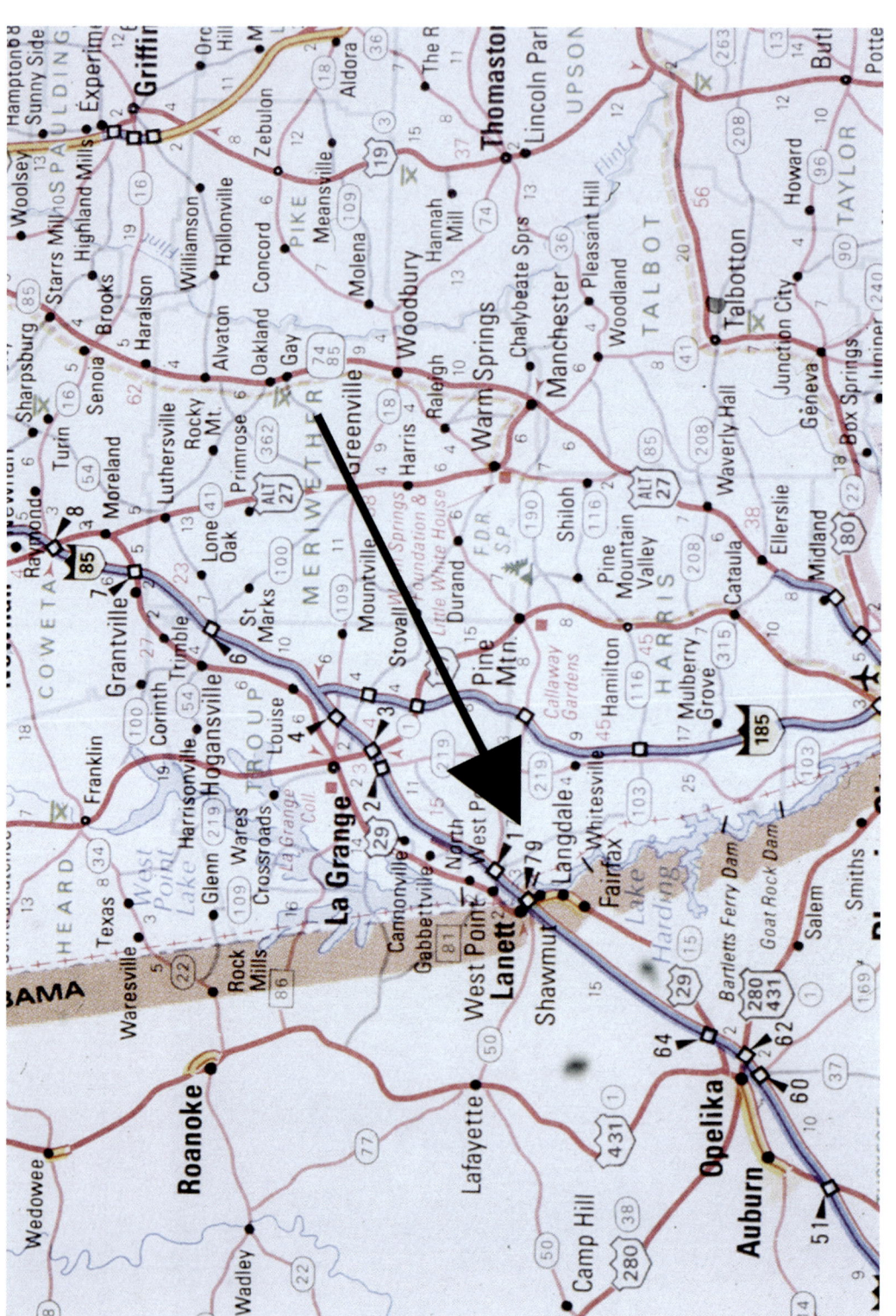

Figure 4.6
Map showing the Crowhop, Georgia area.

through the sieves. Once dried, the samples were visually examined, including examination with a stereomicroscope, and compared with each other. This examination and comparison of the reddish-colored soil from the car and the sample from Crowhop, Georgia, revealed no significant differences (see Figures 4.7–4.9). A density separation method was conducted next, because the quartz and feldspar grains that are usually abundant in a soil sample tend to mask the unique heavy mineral grains. A density separation was conducted on the 45–150 μm fraction of each sample using a modified 30 ml separatory funnel and employing bromoform as the separation liquid (see Figure 4.10). This technique allows the light minerals (specific gravity < 2.89) to be separated from the heavier minerals (specific gravity > 2.89) by settling of the latter. Table 4.1 lists some of the commonly encountered minerals and whether they sink or float in bromoform.

The next step conducted in the analytical scheme involved the examination, analysis, and comparison of the mineral grains contained in the soil samples by

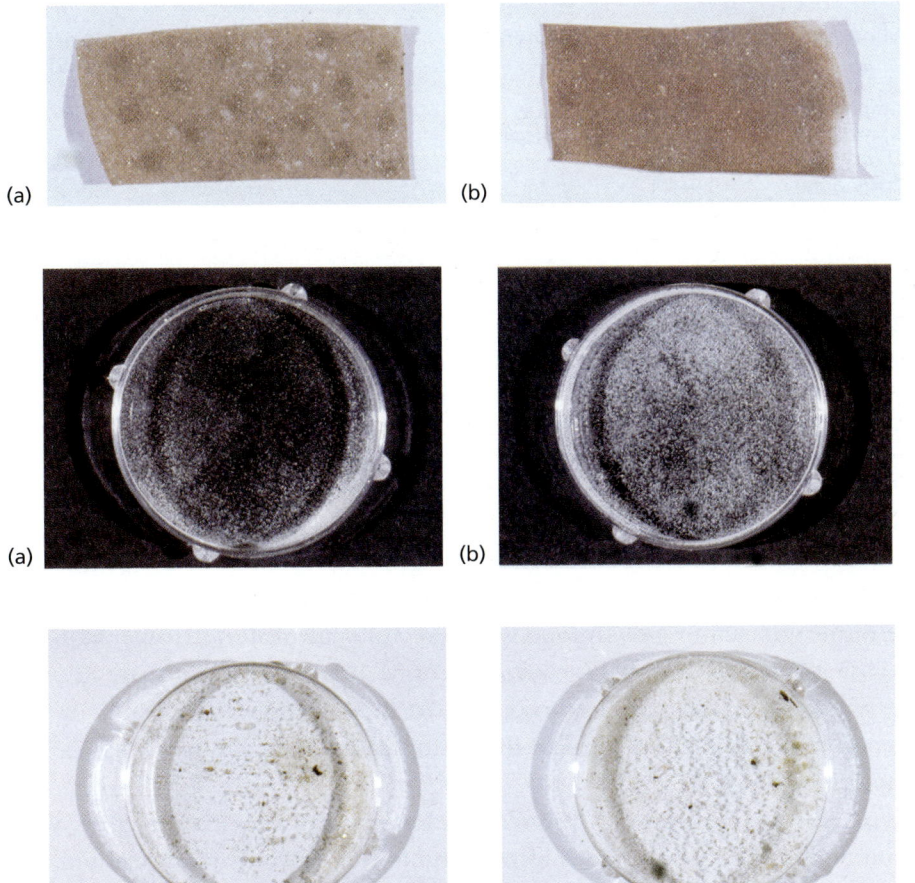

Figure 4.7
Very fine silt/clay fraction (<45 μm) of the reddish soil sample from the car (a) and the sample submitted from Crowhop, Georgia (b) collected on filter paper.

Figure 4.8
The 45–150 μm fraction of the sample from the car (a) and the sample submitted from Crowhop, Georgia (b).

Figure 4.9
The >150 μm fraction of the reddish soil sample from the car (a) and the sample submitted from Crowhop, Georgia (b).

Figure 4.10

Heavy mineral separation using bromoform.

Table 4.1

Gravity of commonly encountered minerals and whether they sink or float in bromoform.

Sink		May sink		Float	
Minerals	Specific gravity	Minerals	Specific gravity	Minerals	Specific gravity
Amphiboles	2.9–3.5	Micas	2.7–3.4	Quartz	2.65
Apatite	3.2	Dolomite	2.8–2.9	Calcite	2.7
Chlorite	2.9–3.0			Feldspars	2.5–2.8
Corundum	3.9–4.1			Gypsum	2.3
Epodote	3.2–3.5				
Garnet	3.8				
Hematite	5.2				
Kyanite	3.6				
Olivine	3.3–3.4				
Pyroxenes	3.3–3.6				
Rutile	4.2–4.3				
Sillimanite	3.2				
Straurolite	3.6–3.8				
Tourmaline	2.9–3.2				
Zircon	4.7				

polarized light microscopy (PLM). Although there are hundreds of minerals, only 30 or so are commonly seen in soil samples. PLM used in conjunction with the dispersion staining technique is a powerful analytical tool available to the forensic trace analyst to characterize, identify, quantify, and compare mineral grains found in soil samples according to their morphological and optical properties. Table 4.2 lists some of the identification characteristics that can be determined by PLM.

For each of the samples, mineral grains from the heavy fractions were mounted on microscope slides using Cargille 1.66 refractive index liquid and grains constituting the light fraction were mounted using Cargille 1.545 liquid. PLM examination and comparison revealed the samples from the car and from

				Table 4.2
Size	Morphology	Surface	Texture	*Characterization and identification properties of mineral grains by PLM.*
Color – Pleochroism	Relief – Dispersion staining colors	Refractive index (indices)	Birefringence	
Sign of elongation	Extinction angle	Twinning	Zoning	
Interference figure	Optic sign	Crystal system	Inclusions	
Alteration(s)	Microchemical test(s)			

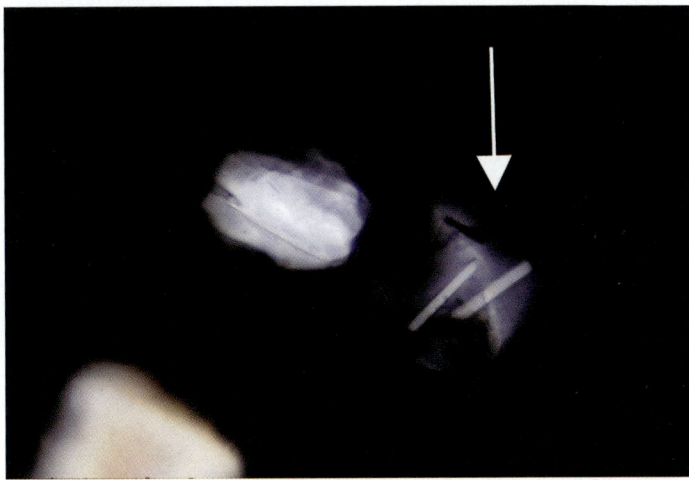

Figure 4.11

Photomicrograph of a typical quartz grain with "rutile" inclusions viewed with crossed polars. Both samples had quartz grains with this characteristic.

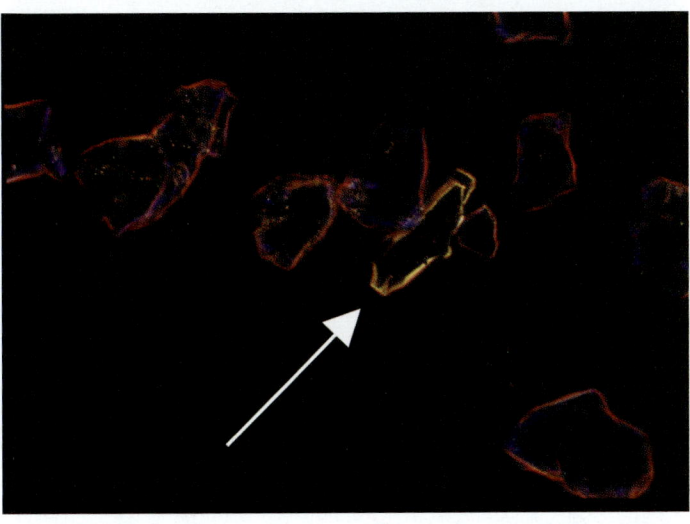

Figure 4.12

Photomicrograph of typical minerals from the light fraction viewed with central stop dispersion staining (no polar). The magenta-colored grains are quartz and the yellow grain is muscovite. Both samples had this composition.

Crowhop, Georgia, to have the same mineral composition, differing slightly (but not significantly) in their relative abundances. Photomicrographs of some of the mineral grains found in the samples are provided in Figures 4.11–4.20.

To confirm that the soil may be from the Crowhop, Georgia area, the author, accompanied by the late Dr. C. J. Rehling, Director Emeritus of the Alabama

Figure 4.13

Photomicrograph of a typical biotite grain from Crowhop, Georgia (left) and from the car (right) viewed with a single polar.

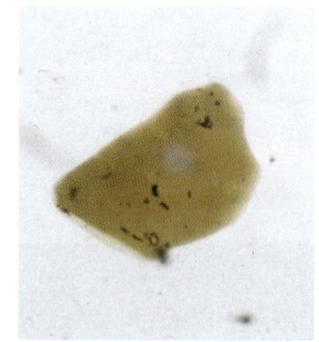
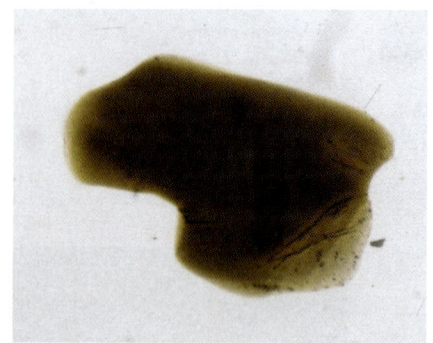

Figure 4.14

Photomicrograph of a typical epodote grain from Crowhop, Georgia (left) and from the car (right) viewed with a single polar.

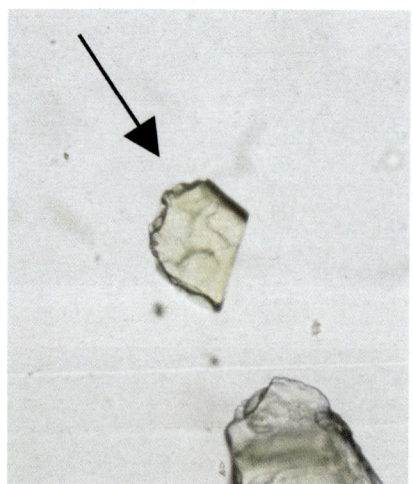

Figure 4.15

Photomicrograph of a typical zircon grain from Crowhop, Georgia (left) and from the car (right) viewed with a single polar.

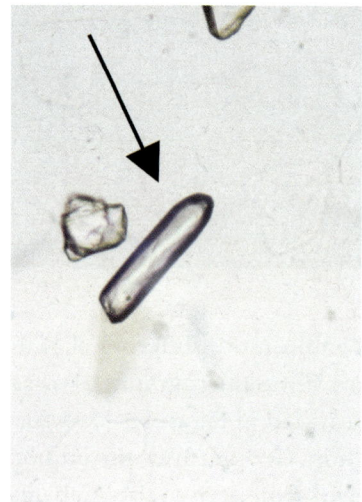
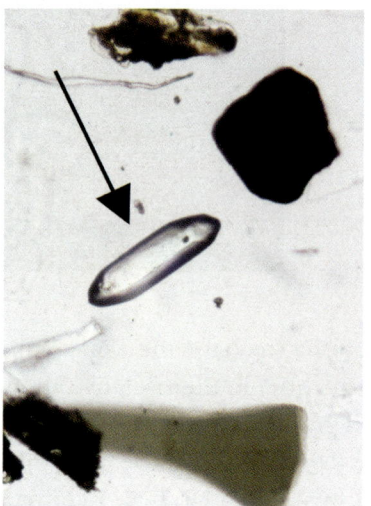

THE VALUE OF SOIL EVIDENCE 117

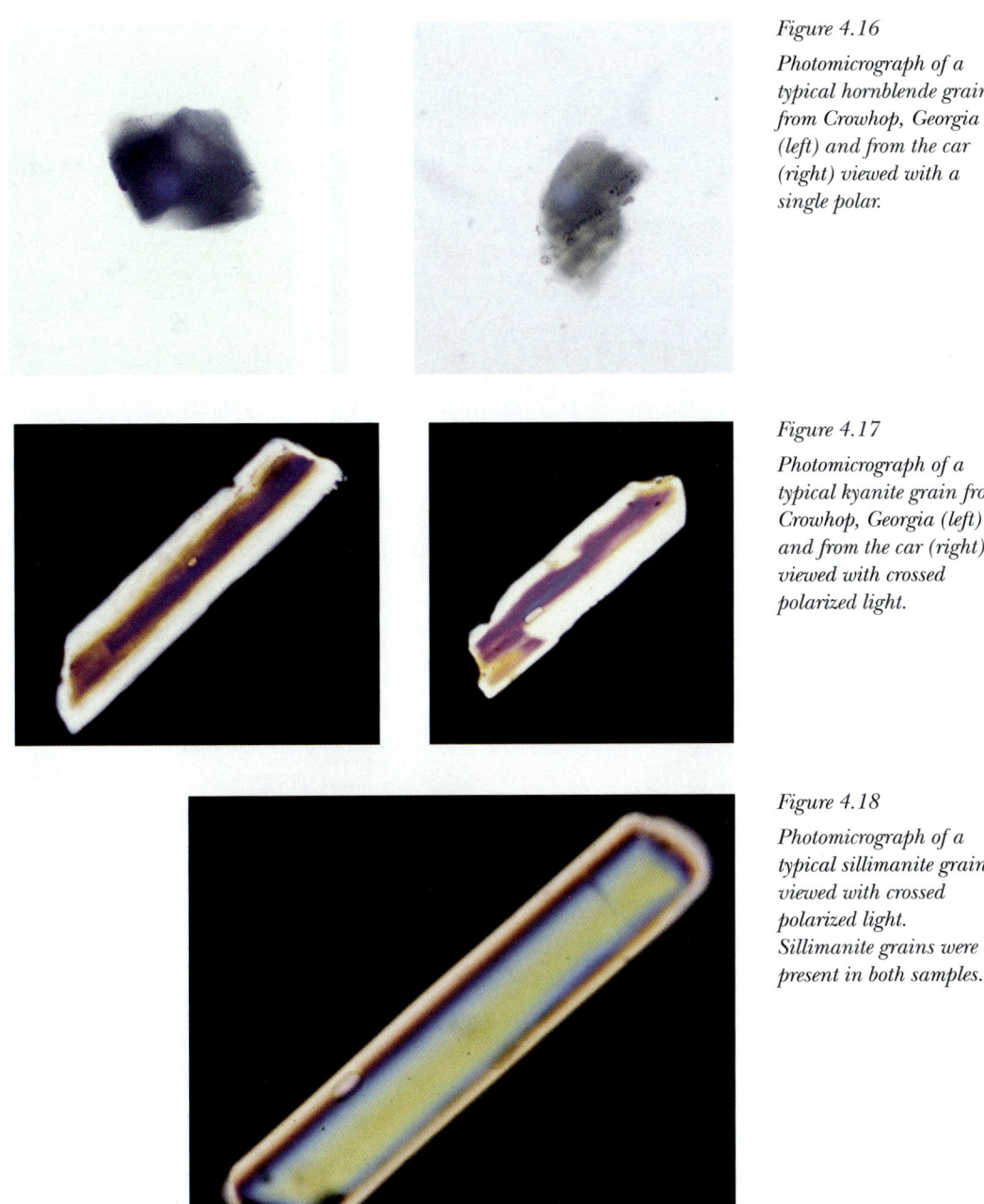

Figure 4.16
Photomicrograph of a typical hornblende grain from Crowhop, Georgia (left) and from the car (right) viewed with a single polar.

Figure 4.17
Photomicrograph of a typical kyanite grain from Crowhop, Georgia (left) and from the car (right) viewed with crossed polarized light.

Figure 4.18
Photomicrograph of a typical sillimanite grain viewed with crossed polarized light. Sillimanite grains were present in both samples.

Department of Forensic Sciences, spent a day collecting additional soil samples along highways and unpaved roads from both sides of the Alabama/Georgia state line. Dr. Rehling was a geologist and provided invaluable insight on the geology of the region. Again, a sample collected from a pipeline right of way

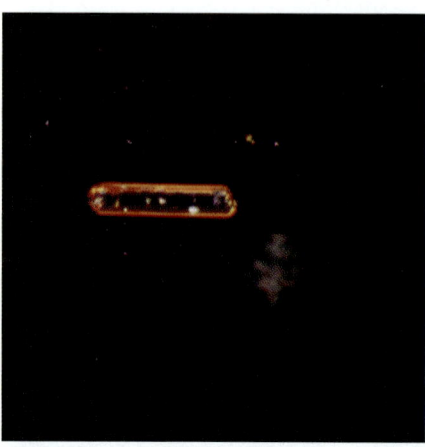
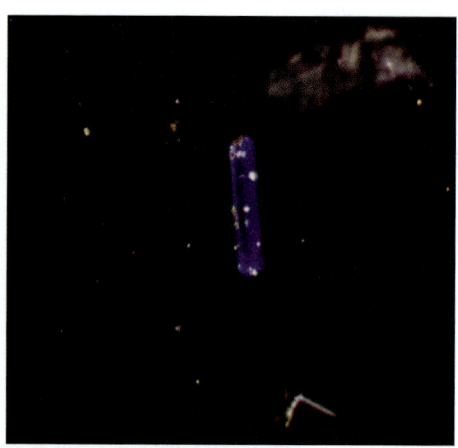

Figure 4.19

Photomicrograph of the sillimanite grain shown in Figure 4.18 viewed with central stop dispersion staining (single polar). Grain length is parallel to polar (left) and perpendicular to polar (right). Both samples contained sillimanite grains.

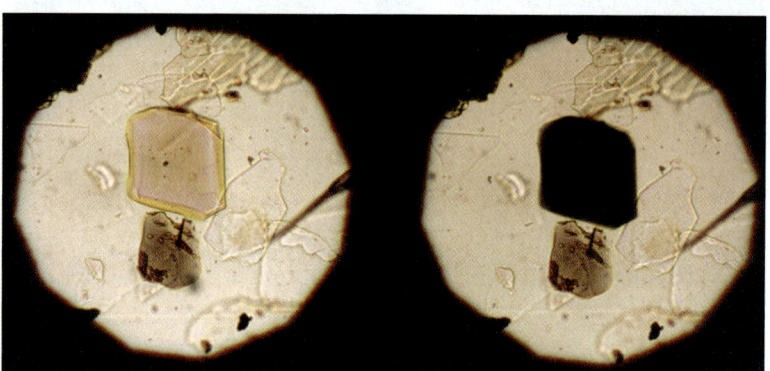

Figure 4.20

Photomicrograph of a typical tourmaline grain (upper grain) showing characteristic pleochroism. The polar is east–west in the left image and north–south in the right image. Tourmaline grains were present in both samples.

along Highway 103 in Georgia could not be excluded as a possible match to the reddish soil using the methods described above.

SUMMARY

Mark Miller was charged with kidnapping and a federal trial was scheduled for November 1983. Murder charges were not filed since a body had not been found. Several days before the trial, the defendant decided to plead guilty in Federal Court on kidnapping charges and to also plead guilty to murder charges in state court, with the sentences to run concurrently and the time served in Federal prison. The acceptance of the guilty pleas was dependent on the defendant showing the authorities where he had buried the body of Alice Redmond.

After pleading guilty in the respective courts, the defendant led the authorities to where he had buried the body. The scene was located down an unpaved road across the state line off Highway 103 in Georgia. The body had been

Figure 4.21
The crime scene off Highway 103 near Crowhop, Georgia.

Figure 4.22
The shallow grave showing the skeletal remains of the victim and articles of clothing found at the crime scene near Crowhop, Georgia.

buried in a shallow grave near a cleared area used for parking for a gun club (see Figures 4.21 and 4.22). This location was within a half-mile from where the author had collected a sample for comparison purposes (see Figure 4.23). Samples from the scene were collected and compared to the samples from the car. Examination and analysis of the samples revealed the soil from the dirt road off Highway 103 leading towards the gun club to be consistent with the reddish soil from the car (see Figure 4.24).

How was the soil evidence significant in this case? First, the recently deposited reddish soil from the car helped establish that it could likely have been in the area across the state line, supporting the story of the accused's sister. Second, it helped to focus the search area in and around the state line. One must wonder

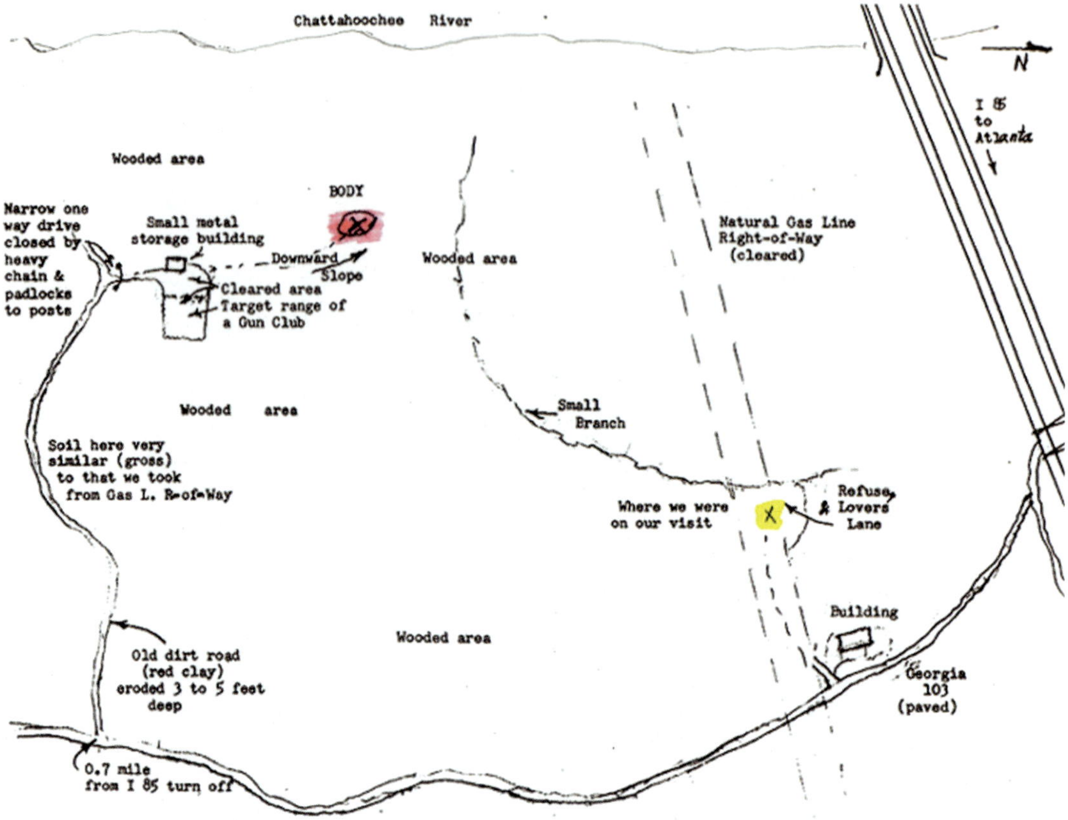

120 TRACE EVIDENCE ANALYSIS

Figure 4.23

Map showing the area where the body was found (highlighted in pink) and the location where the soil sample was taken by the author (highlighted in yellow). The map was drawn by the late Dr. C. J. Rehling, Director Emeritus of the Alabama Department of Forensic Sciences.

Figure 4.24

Soil samples collected in the vicinity of the crime scene near Crowhop, Georgia, and the soil sample from the car. Note the similarity between the soil from the dirt road off Highway 103 (entrance to the crime scene) and the soil from the car.

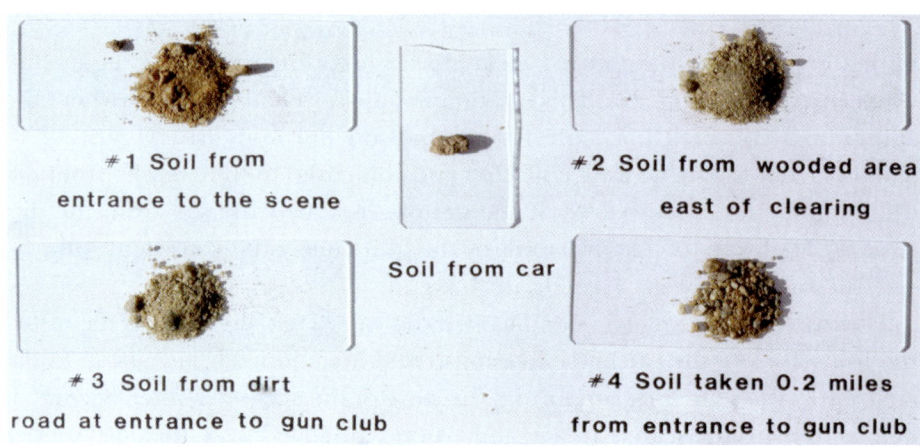

if the suspect had gone to trial and been found guilty on the kidnapping charge, what would have happened if the victim's remains were later found using the soil evidence to help focus the investigation in and around the Crowhop area. Would he have gone to trial a second time on the murder charges with stiffer penalties awaiting him if he had been found guilty?

Some may contend that the conclusions in this case were nothing more than a lucky guess. This author would like to believe that soil evidence is a valuable form of associative physical evidence and does have evidentiary value. I find it puzzling why most forensic laboratories ignore the evidentiary value of soil evidence.

ACKNOWLEDGEMENTS

This chapter is dedicated to the late Dr. Walter C. McCrone who was my mentor and good friend. The cases described within this chapter, as well as many other cases that I have worked throughout my career, would not have been possible without the knowledge "Doc" shared throughout his life with anyone who was willing to listen. Early in my career I was lucky and took my first course from Dr. McCrone where he tirelessly enlightened the class to the analytical power of polarized light microscopy (PLM). To this endeavor, I thank him, and I will try my best to continue to "spread the word". Also, I would like to thank Mr. Randy Boltin and Ms. Lee Brun-Conti for taking the time to review this manuscript.

REFERENCES

Bloss, F.D. (1961) *An Introduction to the Methods of Optical Cyrstallography*. Holt, Rinehart and Winston, New York, NY.

Dana, E.S. and Ford, W.E. (1958) *Dana's Textbook of Mineralogy*. John Wiley and Sons, New York, NY.

Deer, W.A., Howie, R.A. and Zussman, J. (1978) *An Introduction to the Rock Forming Minerals*. Longman Group Limited, London, England.

Graves, W.J. (1979) "A Mineralogical Soil Classification Technique for the Forensic Scientist," *JFSCA*, 24, 323–338.

Heinrich, E.W. (1965) *Microscopic Identification of Minerals*. McGraw-Hill, New York, NY.

Mange, M.A. and Maurer, H.F.W. (1992) *Heavy Minerals in Colour*. Chapman & Hall, New York, NY.

McCrone, W.C. (1982) "Soil Comparison and Identification of Constituents," *Microscope*, 30, 17–25.

McCrone, W.C., McCrone, L.B. and Delly, J.G. (1978) *Polarized Light Microscopy*. Ann Arbor Science Publication, Ann Arbor, MI.

McCrone, W.C. (1992) "Forensic Soil Examination," *Microscope*, 40, 109–121.

McPhee, J. (1996) "Grounds for Murder," *The New Yorker*, January, 44–69.

Murray, R.C. (1982) "Forensic examination of soil," in *Forensic Science Handbook*, ed. R. Saferstein. Prince-Hall, Inc., Englewood, NJ.

Murray, R.C. (2000) "Devils in the Details – The Science of Forensic Geology," *Geotimes* (www.geotimes.org), February, 14–17.

Palenik, S.J. (1982) "Microscopic Trace Evidence – The Overlooked Clue," *Microscope*, 30, 163–169.

Petraco, N. (1994) "Microscopic Examination of Mineral Grains in Forensic Soil Analysis: Part 1," *American Laboratory*, April, 35–40.

Petraco, N. (1994) "Microscopic Examination of Mineral Grains in Forensic Soil Analysis: Part 2," *American Laboratory*, September, 33–35.

Stoiber, R.E. and Morse, S.A. (1981) *Microscopic Identification of Crystals*. Robert E. Krieger Publishing Co, New York, NY.

CHAPTER 5

THE IMPORTANCE OF TRACE EVIDENCE

Harold Deadman
George Washington University, Washington, DC

INTRODUCTION

A missing person, a ransom request, a payoff, and a man is apprehended with the ransom money. After being arrested the subject gives five different stories as to how he received the money. It would appear there is an open and shut case of kidnapping. The missing person is found beaten and shot to death in a field and the suspect is identified as the forger of one of the victims' checks. Do we now have an open and shut case of kidnapping and murder? Many men have been convicted on much less evidence. Over the last 10 years post conviction DNA testing has exposed many instances where innocent men have been wrongly convicted. Many, if not all, of these erroneous convictions were based on little evidence, erroneous evidence and, in some cases, physical evidence that was incorrectly analyzed. This case was different. There seemed to be plenty of strong evidence of guilt. The case began in 1985, but even with compelling evidence, was not resolved until July 15, 1994. Is it over? Perhaps, but as it often seems with the legal system, justice is a moving target.

This chapter deals with the analysis and significance of hair and textile fiber evidence in a criminal trial, discussing the many aspects of hair and fiber analysis and how they were used in a case of kidnapping and murder. The case and the trial occurred in South Dakota in 1985. Hair and fiber evidence was important evidence at the trial demonstrating an extremely strong connection between the victim's body and the trunk of an automobile used by the subject. This case is similar to many cases where, although hair and/or fiber evidence was important evidence at the criminal trial, it was only one of many meaningful pieces of evidence presented by the prosecution. There have only been a few cases where hair evidence and/or fiber evidence have been the essential evidence in a case. It is most often used to support other evidence, to supplement and complement other evidence and that was the situation here. There was considerable evidence to strongly link the subject to the crime but the hair and fiber evidence, in addition to strengthening the link between victim and suspect, also helped tell a story that could not be told by the other evidence. It will discuss the actual evidence introduced in a kidnapping and murder trial and the significance of that evidence. There were other interesting aspects to the case, some scientific and some legal, as the defense attempted to prevail in the appeal process when they could not in the trial. Some of the legal aspects and legal decisions that came out of the trial will also be reviewed; as is often the case, the legal challenges and proceedings after the trial took much longer than the investigation and trial. This case is a good example of what occurs in a highly contested criminal trial, where much of the prosecution evidence is challenged during and after the trial.

EVIDENCE

Evidence in a criminal case is something that makes certain contested issues more evident, and allows one to get closer to the truth of a matter. It is something that relates to or attempts to support an issue in question and in a criminal trial is almost always presented to a jury of one's peers. At least two sides are presented in a trial, especially in a homicide. Rarely does anyone, at least anyone who is willing to talk, know exactly what happened during the commission of a homicide. The prosecutor has a hypothesis about what occurred during the crime and should have strong evidence that supports his theory of the crime. The prosecution presents its evidence to a jury and attempts to convince the jury that not only is the evidence valid but that it is proof beyond a reasonable doubt that the prosecutor's hypothesis is correct; reasonable doubt is the standard under which a criminal jury must operate. The defense also has a conjecture about the circumstances and can take several approaches in presenting their evidence to the jury. Often they will attempt to discredit the prosecutor's evidence or discredit the witnesses used by the prosecutor. They can also accept the prosecutor's evidence but provide the jury with different interpretations of that evidence. Alternatively, the defense may have no story to tell and simply contest or dispute the prosecutor's hypothesis to try and show that it is wrong or that the evidence is not sufficient for the jury to draw the conclusions that the prosecutor desires. The jury, after hearing both sides of the case, must decide if the prosecutor's evidence is sufficient for conviction. To do this they must first decide what reasonable doubt means to them and apply their definition of reasonable doubt to the evidence. In the case of scientific evidence, the jury will normally rely on expert witnesses to explain what the evidence means and their job is to accept or reject the expert's opinion testimony. The jury will often have a difficult time in determining what or who to believe, especially if conflicting and opposite interpretations of the scientific evidence are presented by different experts.

What is required to convict someone of a crime, to convince a jury beyond a reasonable doubt? Reasonable doubt is a term often used, but not easily defined. There is no single definition for proof beyond a reasonable doubt. However, most courts describe it as being less than the absence of doubt and more than a probability. Another definition, given by a judge instructing a jury, stated that reasonable doubt is not mere possible doubt, because everything related to human affairs, and depending on moral evidence is open to some possible or imaginary doubt. It is that state of the case, which, after the entire comparison and consideration of all the evidence, leaves the minds of the jurors in that condition that they cannot say they feel an abiding conviction to a moral certainty, of the truth of the charge. Based on a review of numerous erroneous convictions that have been reported over the past 10 years, often it does not take much. Other commentators have stated that reasonable doubt is a doubt based upon reason and common sense. It is a fair doubt and not a vague, captious or imaginary doubt, but a doubt growing out of the evidence, or lack of evidence, or the unsatisfactory nature of the evidence in the case. When one looks at these definitions of reasonable doubt, one can see why juries can have problems and sometimes come to the wrong decision.

But there have been cases where the evidence has been overwhelmingly in favor of guilt and yet the result is no conviction. There are obviously other reasons for a jury verdict other than the evidence in a case but the forensic scientist, however, must be concerned only with correctly analyzing and interpreting what the evidence means, present their evidence to a jury, and defend their results and conclusions.

The points of law and legal arguments used by lawyers at a trial can have an effect on the evidence and this case was no exception; judges can also influence the interpretation on forensic science in the courtroom and even the appeal process. As one would expect for a case that went on for almost ten years, the case record is extensive. There were approximately 2800 pp. of trial transcript, 193 trial exhibits, several motion hearings, 220 pp. of briefings, and three supplemental briefs.

THE FORENSIC SCIENTIST

The role of the forensic scientist in the criminal justice system is complex and varied. Some maintain the forensic scientist's job is simply to conduct experiments and testify about the results: This is a simplistic view. One of their most important responsibilities is to conduct discriminating test procedures so as to get the right answer in determining which of the items being tested match and do not match.[1] After different samples have been shown to exhibit correspondence in characteristics, the analysis is far from over. The significance or evidential value of the

[1] The term *match* is used throughout this chapter for simplicity purposes. There are many in the legal community and some in the forensic community who do not like the word "match" and argue that a jury can misinterpret it. Any word or term used in a report or in court can be misinterpreted, so an analyst must define terminology, regardless of the terminology used, and especially if it is used in court. A "match" is nothing more than a statement that corresponding properties were found in the evidence being compared and that no significant differences were observed. "Match," however, is typically not used in reports or in court testimony by hair and/or fiber examiners.

matching results must be determined and then reported, along with their significance, in a clear and understandable fashion. After educating and explaining their evidence and the significance of their results in the courtroom, forensic scientists often have to defend their testimony against inquiry and criticism during cross-examination. Forensic science is one of the few occupations where scientists can use proper testing procedures, get the correct answer, interpret the results correctly, and presenting their findings clearly but still undergo vigorous attack and criticism. What makes a forensic scientist's job more complicated is a continuing debate regarding proper analysis, proper reporting, acceptable testimony, accurate conclusions, and the requirements necessary to support the conclusions. There are often cases where expert witnesses on the opposite sides of a case will present completely contradictory results, different interpretations, and different conclusions. In addition, opposing experts will downplay or minimize the evidential value or significance as set forth by the other expert. The ability to educate and explain the results and the significance of the evidence is extremely important. If the world's greatest scientist was an analyst in a forensic case and conducted the best possible analysis, the results and conclusions would amount to nothing if they were not presented in a logical, concise and simple but convincing way.

THE IMPACT OF DNA TESTING

The last decade has brought about profound changes in forensic science in large part to the development of DNA testing. DNA technology has had a tremendous impact on the solution of crimes of violence especially those involving sexual assault: For once the source of a biological sample is known. DNA testing is so sensitive and powerful that the chance of matching DNA to the wrong person (a coincidental match) is in most cases so small, it can be ignored. Another important aspect of forensic DNA analysis is that the interpretation of DNA testing results is straightforward and it is difficult to get a wrong answer; there is little subjective analysis involved in the interpretation of DNA typing results. DNA protocols have been validated, are monitored, and have been extensively tested over the past 15 years. The great sensitivity of the current methods usually means that there is almost always sample remaining for additional testing.

DNA testing also has had a considerable influence on other forensic science fields and on the operation and functioning of a crime laboratory. Because of DNA testing and the creation of DNA databases, the funding and staffing of crime laboratories has increased significantly. But these new resources have been directed toward DNA testing and, therefore, not to other parts of the crime laboratory. This has been especially true with the part of the laboratory that conducts hair and fiber analysis. Hairs and/or fibers are often extremely important in helping convict those committing violent crimes; in fact, hair and fiber evidence was considered to be much more probative than the classical serology testing procedures that were used prior to DNA. But because DNA testing produces powerful evidence for identification purposes, and has been demonstrated to be reliable and produce reliable results, many crime laboratory directors believe that the analysis of trace evidence, especially hair and fiber analysis, is not necessary. Some laboratories have either reduced their commitment to hair and fiber testing or, in some cities and states, eliminated it completely.

DNA has also impacted greatly on forensic science in the courtroom. Because DNA testing became such powerful evidence and because it was new and was so damaging to defendants in a criminal trial, there were numerous extensive attacks by defense attorneys in the courtroom to try and defeat it. To accomplish this purpose, expert witnesses hired by the defense were used in thousands of cases to attack and criticize the DNA evidence or present different interpretations of the evidence. Prior to the introduction of DNA testing, defense experts, at least in trace evidence, were not often seen in the courtroom, but this has changed. It is likely that all fields of forensic science are going to experience more frequent and continuing attacks and criticisms in the courtroom and encounter defense experts more often. Admissibility of many types of forensic evidence has become an issue or will become an issue in the future.

The impact of DNA testing has also spurred the development of Scientific Working Groups, placed great emphasis on quality assurance in the laboratory, encouraged the involvement of academics in forensic science, and through the working groups have created technical committees to evaluate methods and procedures. With no question, this focus on and scrutiny of DNA has benefited the entire field.

TRACE EVIDENCE

Trace evidence is evidence that is transferred in small amounts when there is contact between individuals and/or objects. When contact occurs, material will be transferred in amounts that can be recovered and analyzed. Trace evidence includes materials such as hairs, fibers, glass, paint chips and smears, soil, and anything else that can be easily transferred in small amounts. Biological evidence, even though often transferred in trace amounts, is not considered to be in the same category as hairs, fibers, and other trace materials. Trace evidence typically is used to link together people, places, and physical objects and can often result in extremely strong evidence of association. Because it provides indirect information about the circumstances that took place before, during, or after criminal activity, trace evidence is considered circumstantial; eye witness testimony is direct information. Circumstantial evidence is often criticized in the courtroom but many believe that it is more powerful, more reliable, and provides more valid information about crime events than eyewitness identification. Most evidence presented at a trial, however, is circumstantial evidence and most homicide trials rely on circumstantial evidence. In fact, trace evidence can tell us many details about the actual events that occurred during the crime and functions in many cases as a "mute witness".

Hair and fiber evidence are arguably two of the most important types of the trace evidence. They are obviously important in most violent crimes such as rape, assault, and many homicides, where there is often considerable contact between the perpetrator and victim, to link together the victim and/or the victims' environment with the subject or the subjects' environment. As an example, consider a body that has been moved from a crime scene. This is the ideal type of case to look for trace evidence. A moved body has to be transported from a crime scene, usually with some type of vehicle. There is a good chance that the victim will have left behind trace materials, such as hairs at the crime scene and in the transport vehicle, and even more likely to have picked up trace materials, especially fibers from the crime scene and the transporting vehicle. Even in the age of DNA testing, a dumped body case demands that a thorough examination take place for trace evidence, on clothing, on skin surfaces, and in the hair of the victim. In addition to providing meaningful evidence at a trial, hair and fiber evidence can serve as an investigative aide. For example, it can provide information about the crime scene, about the vehicle used to transport a victim, and about the race and appearance of the perpetrator.

THE EXPERT WITNESS

A forensic scientist testifying about some type of physical evidence is testifying as an expert and expert witnesses are treated differently than other witnesses in a courtroom: They are allowed to give opinions and they can rely on information other than what they know from their own senses. Normally, regular witnesses providing evidence must testify only about facts, what they have seen, heard, or have otherwise witnessed through their senses. In many situations, the facts may be such that only an expert is able to understand and correctly interpret the evidence. In these situations, the special knowledge and training of the expert enable him or her to explain to the jury what the layperson would completely fail to grasp. The principal role of the expert in rendering opinions is to aid the jury (or judge) in determining what the evidence means. The rules under which an expert operates depend on the jurisdiction, the court, the judge, the subject of the testimony, and the actual evidence. In Federal Court and in many state courts the expert witness is allowed to testify under Rule 702 of the Federal Rules of Evidence which states:

> if scientific, technical, or other specialized knowledge will assist the trier of fact to understand the evidence or to determine a fact in issue, a witness qualified as an expert by knowledge, skill, training, or education, may testify thereto in the form of an opinion or otherwise.

But there may be a problem with expert testimony: when juries hear conflicting testimony from expert witnesses for the prosecution and the defense. This problem is magnified when both witnesses appear to have good credentials but different opinions about the evidence.

HAIR AND FIBER EVIDENCE

The emphasis of this chapter will be on two types of trace evidence: human hair and textile fibers. Hair and fiber evidence can:

- link a suspect to a crime scene or a victim,
- establish a sequence of events,
- link a murder weapon with a victim or suspect,
- help to corroborate a victim's account of circumstances surrounding an assault,
- provide leads about murder victim's surroundings at the time of the murder,
- link together a number of different (sometimes apparently unrelated criminal activities),
- establish a high likelihood that contact has taken place between a victim, a suspect and/or a crime scene.

Although it is true that DNA testing can accomplish much of the above when biological evidence is present, there are cases where biological evidence will not be recovered and trace evidence needs to be considered. Hair and fiber analysis can be quite complicated because it deals with:

- evidence recovery outside the laboratory,
- development of information about the case,
- evidence recovery inside the laboratory,
- evidence handling and preparation for analysis inside the laboratory,
- the screening of hairs and fibers,
- microscopical characterization and comparison of probative hairs and fibers,
- other methods of analysis including chemical, instrument and/or biological testing,
- the assessment of evidential value – what does the evidence mean?
- reporting of the laboratory results,
- direct testimony about the results and conclusions,
- defense of one's results and conclusions.

A good examiner requires many different skills and must have education, training, experience, knowledge, and skill in each of the above areas. An examiner who examines hairs will also examine fibers, and vice versa. There are several reasons for this. Probably most important is the fact that hair and fiber evidence is going to be collected at the same time and microscopical techniques, especially the use of the bright field transmitted light comparison microscope, are important tools in both types of analysis. There are, however, many differences between the analyses of the two types of evidence. Until the advent of DNA analysis, the only useful method for examining and comparing human hair was with the comparison microscope. An examination involved the side-by-side comparison of a hair of unknown origin, or what is called a questioned hair, with numerous hairs from a particular source or what is called a known hair sample. The examination is difficult and relies on the judgment and, most importantly, the skill of the analyst. The comparison microscope is also an important tool for fiber comparison but many other methods of characterization and comparison are brought to bear on fiber than on hairs.

FORENSIC HAIR ANALYSIS

Human hair comparison is a technique that is not well understood, even by some individuals in a crime laboratory, as hair evidence is different than any other type of evidence. A human hair has many structural features and characteristics that can provide information about the hairs and be used in the comparison of a hair of unknown origin to hairs from a particular person. These comparisons require the use of microscopes and hairs are typically examined with research quality microscopes at magnifications up to 400×. A comparison microscope consists of two-research quality microscopes that are bridged together and typically costs from $10,000 for a very basic system to over $50,000 for the more sophisticated systems.

There are a number of reasons why forensic hair analysis is different from other forensic examinations. Hair is continually growing and its characteristics can vary along its length as it grows. For example, a head hair that is 6 in. in length has been growing for approximately 15 months. It will have a tip end that has been subjected to hundreds of days of sunlight, shampooing, brushing, combing, etc. The portion of the hair that has just emerged from the shaft will have been subjected to none of these environmental or cosmetic conditions. Differences, sometimes large, will exist in the different sections of the hair. Hairs also go through growth cycles: a hair is either actively growing or in a resting stage where the hair is on the verge of being sloughed off the body. There are differences in the characteristics of hairs in different growth cycles. As a person ages, hairs also change. This is obvious in the color variation that exists on a person's head as they age and the loss of pigmentation causes some hairs to become white or gray but other more subtle differences also occur because of aging or because of the location of the hair on the scalp. The hairs in the different body areas show considerable differences requiring that hairs from the same body area be compared. This can complicate matters when one has a transitional hair, a hair that comes from an area where you might expect the characteristics of two different body areas to be present, such as in a hair from a man's sideburns. Often the difference between hairs from different individuals is great but interpersonal differences can also be small and very subtle. No two hairs, even from the same person, are going to be exactly alike and assessing what differences are significant and are not significant between hairs being compared can be difficult.

It was mentioned earlier that hair growth could cause differences in characteristics between different portions of the same hair. In addition, a hair is three-dimensional and its cross section is not often perfectly circular so its appearance under a microscope will depend on how the hair is sitting on the microscope slide. Considerable variation in the features that occur within the hairs from particular body area from one person also may exist. This intrapersonal variation also sets hairs apart from many other type of trace evidence. Because of this variation, numerous hairs, usually between 20 and 100 are collected for a known sample from a particular body region to be used in the comparison process (Bisbing, 2002). This methodology – collecting type specimens for detailed examination and comparison – demonstrates the origins of forensic hair examinations in comparative biology and taxonomy.

Even though forensic hair examinations and comparison have been used in the courtroom for many years, the field of forensic hair analysis has had many problems and has its detractors. In England, for example, forensic hair analysis based on microscopical analysis is not considered a useful technique. The following represents an English forensic scientist's concerns about forensic hair comparisons:

- Acquisition of hair evidence is often lengthy, labor intensive and subjective.
- Confidence in hair examination is often low.
- The same methods and criteria have been used for the past 50 years.

Another concern or problem with forensic hair analysis that has come to light over the past several years is the most serious obstacle facing the use of microscopical hair analysis. Forensic hair comparisons have been involved in incorrect associations in actual criminal cases. These mistaken associations have occurred primarily in rape cases where matching hair evidence was used as evidence of association between the suspect and victim at trials that resulted in convictions. In these cases, the convictions were shown to have been wrong by subsequent DNA analysis of biological evidence. Biological evidence that was thought to have originated from the perpetrator was found to be from someone other than the man who was convicted. There are two reasons for mistakes like this. First, the matching hairs originated from two different individuals but exhibited the same microscopic characteristics by coincidence, or, second, the examiner did not recognize meaningful differences that should have precluded a match determination. When a hair has been shown to not originate from a person to whom it was microscopically matched, it is critical to determine why the match occurred. Unfortunately in these cases, the next step to determine if the examiner or the mismatched hairs were the source of the error is not taken. The incorrect inclusion type of error (also called a type II error and the type of error that occurred in the erroneous convictions) is particularly problematic (Bisbing, 2002). The other type of error, an incorrect exclusion, where a hair is incorrectly associated to or incorrectly excluded from originating from a person can be serious but does not result in evidence that can be used against an innocent person.

As pointed out earlier, forensic laboratories are often under funded, especially local and state laboratories. The influx of DNA money did not help matters much for disciplines outside of DNA. With limited funding, a laboratory typically did not have the staff or resources to specialize in fields like hair analysis or have the money to buy expensive microscopical equipment, such as a high quality comparison microscope. Crime laboratory management often does not understand the process of hair comparisons and those hair examinations are difficult and time-consuming, and that it can be easy, especially for the inexperienced examiner, to make a mistake. Specialization and extensive training is extremely important in any field but especially in a field whose analysis is less objective than most other analyses. Also, few formalized training programs exist for hair examiners.

Another issue is that, historically, most witnesses with some experience with human hair would be allowed to testify as an expert. Anthropologists, cosmetologists, and dermatologists have been allowed to testify in court even though these individuals have never conducted a microscopical forensic hair comparison. Their ability to correctly conduct hair comparisons has never been tested. It is also likely that judges do not understand hair analysis and they have no way to evaluate the ability or the knowledge of a witness. For that matter, laboratory management, unless they have extensive hair-training programs or have been hair examiners themselves, also have no way to evaluate ability. Lately, however, qualification in court is not as simple a task as it used to be.

Significance is another concern in forensic hair comparisons. One important measure of the value of any type of associative evidence is the probability of getting the matching results in a case by chance or coincidence. Because hairs from different individuals generally have different characteristics and hair characteristics can range from being very common to very rare, the value of hair evidence will also vary. Assuming a match is correct, the more uncommon the characteristics of a hair involved in the match, the smaller the probability of a coincidental match and the greater the value of the hair evidence.

Perhaps, the most serious issue and limitation in forensic hair analysis is assessing a hair examiner's ability. Performance (in getting the correct answer) should be the key with any type of analysis where little if any quantitative data is generated for comparison. One cannot easily use actual casework to measure performance because one is never absolutely sure of the correct answer. Performance must be measured with tests that mimic casework but tests where the correct answer is known (but obviously not by the person being tested). Arguably, the only way to develop properly-trained forensic hair examiners is to regularly train hair examiners using matching tests that measure and monitor performance. Training for a hair examiner should take at least 1 year and should include numerous matching tests graded so that the test's difficulty increases throughout the training period. Even after becoming an examiner one should undergo regular proficiency testing to measure performance. Numerous testing has not been part of a typical training program in most laboratories and is one of the reason for the poor performance of many hair examiners. In addition there are presently no adequate proficiency tests programs available from outside vendors. This should not prevent extensive in-house testing of hair examiners by a laboratory, however, as proficiency test that mimic casework are easy to produce. And, of course, it is the laboratory that is responsible for producing and using competent and qualified analysts. Many laboratories rely on an apprenticeship approach to training hair examiners. This does simplify many parts of the training process but cannot replace vigorous and regular testing of trainees with practical tests that mimic casework.

Because of the difficulties and limitations discussed above, it is also important that examiners set high standards for a hair match. This is true no matter what significance an examiner assigns to the hair match, but especially if the examiner is going to argue that the hair match is the basis for a strong association. As pointed out by Gaudette, the statement "If in doubt throw it out" should be the motto of all hair examiners (Gaudette, 1999).

One of the criticisms of forensic hair comparisons is that it is "subjective". It needs to be pointed out, however, that a hair comparison is not subjective in the same way as a subjective assessment. The point-by-point microscopical comparison of hairs is more exacting and involved than merely asking, "Are these two things similar?" A hair examination involves a comparison of features that two properly trained examiners should be able to agree on even if they do not agree on what to call or how to describe those characteristics. The examination of polymer surfaces involves subjective impressions and these contain a great deal of information – and the recognition of these features is nearly instantaneous by an experienced examiner. Storing and transferring this information, however, is not possible in a quantitative sense. This fact does not detract from the reality that subjective processes hold many advantages. This statement also holds for forensic hair comparisons based on microscopic characteristics.

What is the future of forensic hair analysis? Many believe it is a combination of both microscopical comparison and DNA analysis. Hair is biological and contains DNA. Unfortunately, the DNA testing procedures developed and used over the past 15 years have not generally been useful in hair analysis. Until recently, most DNA testing analyzed nuclear DNA that originated in chromosomes present in the nucleus of cells. Nuclear DNA is available for testing in some situations with hairs that have been forcibly removed from the body. If a hair is actively growing (which most are), when the hair is pulled out, standard DNA testing procedures can be used with the hair root to obtain the same type of results as with blood and semen. Forcibly removed hairs can also have tissue adhering to the hair root and shaft from which large amounts of nuclear DNA can be obtained. Many hairs, however, that are recovered in criminal cases are not forcibly removed but are hairs that have been shed naturally by the body. These hairs often fall out and are resting on a person's clothing and will be easily lost or transferred upon contact. It is usually not possible to recover sufficient nuclear DNA from a shed hair to use any of the standard methods used routinely on blood, semen and saliva stains.

There is, however, another type of DNA that can be recovered from shed hairs and used for comparison purposes. Mitochondrial DNA can be extracted in sufficient quantity from hairs including the hair shaft and analyzed (Wilson *et al.*, 1995). The testing procedures are relatively new, more complicated, and the evidential value is not as powerful as with current nuclear DNA methods, but the combination of microscopical analysis and the mitochondrial DNA testing can produce powerful evidence (Houck and Budowle, 2002). In addition, the mitochondrial DNA also serves to validate the microscopical comparison, as it is a very discriminating procedure, detecting somewhere around 95% of the incorrect associations made by a hair examiner. A paper that earlier pointed out the negative attitude about hair comparisons in England contained the following comment about hair analysis: "What is clear, however, is that a combination of mitochondrial DNA analysis and the added confidence this would bring to microscopical hair examination would make this a powerful tool in this rather neglected area". New methods of mitochondrial DNA analysis are being developed and show great promise in reducing analysis times, cost, and sample size (Devaney *et al.*, 2000; Holinski-Feder *et al.*, 2001; Liu *et al.*, 1998; Oldenburg *et al.*, 2001; Roberts *et al.*, 2001; Underhill *et al.*, 1997) some of which have already been applied to identifying the victims at the World Trade Center terrorist attacks on 11 September (Miller, 2002).

It will also be some times before more than a few public laboratories actually conduct mitochondrial DNA analysis. Because of the current complexity, high cost and lack of public laboratories involved in mitochondrial DNA typing, it is likely that only those hairs that are probative and are likely to have originated from a particular person should be tested. The significance of a mitochondrial DNA match depends on the size of the database of mitochondrial DNA types that is available for comparison. If a mitochondrial DNA match is made, the number of individuals that have matching DNA types in a mitochondrial DNA database determines the significance of the match. Mitochondrial DNA databases are constantly being increased in size and a larger database will provide a more accurate estimate of the frequency of a particular DNA type in the general population. But even with larger databases, the combination of DNA testing and microscopical analysis should result in much stronger evidence than either technique alone.

In addition to the comparison of hairs, there are a number of determinations that can be made during a microscopical hair examination. It can be determined if the hair is of human or animal in origin. If animal, it is often possible to identify the family or even species of animal. If human, it is possible to estimate the race of the hair contributor and identify the body area from which the hair originated. Race estimation is treated in a somewhat simplistic way in that most hairs are placed into three categories: Caucasoid, Negroid, or Mongoloid (Asian and Native American) or in a fourth category of mixed racial origin (Houck and Koff, 2000). It is also possible to determine if the hair has been chemically treated, if it has been damaged and how, and if the hair may have been forcibly removed. While these determinations are helpful and can provide lead information to the investigators, the most important aspect of a hair analysis is that certain hairs of unknown origin can be compared with hairs from a particular person. The goal of these comparisons is to identify the source of hair evidence. The human hairs that are usually used in the comparison process and which have value for identification purposes are head and pubic hairs. These are the type of hairs that have been studied extensively and which show considerable variability throughout the general population.

FORENSIC FIBER ANALYSIS

In the analysis of textile fibers, it is possible to identify the polymer class of the fiber and, in some instances, identify the type of textile material from which the fiber originated. But as with hair comparisons the most important aspect of fiber analysis is the ability to compare fibers of unknown origin with fibers from a particular source to try and determine if the known and unknown could have had a common source. As mentioned earlier, forensic hair analysis and forensic fiber analysis are similar in some respects and microscopical analysis is a necessity with both types of evidence, especially the use of the comparison microscope. Forensic fiber analysis, however, is different in many respects from forensic hair analysis. In contrast to hair analysis, there are numerous established procedures that exist for the characterization and comparison of textile fibers. The methods of analysis and comparison are not novel and include objective procedures so there should be no difficulty in the admission of fiber evidence in the courtroom; some of these methods are used by the textile industry in the quality assessment of their products. The available microscopical methods are very discriminating, easy to use, and the tremendous variability of textile fibers means that most fibers will not be common (Roux and Champod, 1990). It is much easier to obtain correct matches in fiber analysis than forensic hair analysis, if the proper procedures are used. This does not mean that mistakes cannot be made in a forensic fiber examination: An examiner still needs to be experienced, set high standards for matching, and use proper protocols and equipment.

It is important that a comparison microscope is used for the forensic fiber comparison process. However, in forensic fiber analysis the comparison microscope is supplemented with numerous other microscopical, instrumental, and chemical procedures. With fibers, there are some parts of the process that are subjective but, unlike hairs, the comparison process produces objective data to be compared. Central to both hair and textile fiber evidence is a concern that is common to all trace evidence but especially problematic for textile fibers: the potential for contamination. The recovery process, both in the field and in the laboratory, can be complicated and there are many chances for contamination to occur if proper procedures are not used and care in processing is not observed. Contamination is the central concern, so much so that the crime scene and laboratory recovery of fibers must also be the primary concern of the forensic laboratory.

EVIDENTIAL VALUE

After a hair is matched to a particular person or a fiber is matched to fibers in the composition of a particular textile material, the analysis is far from being finished. It is necessary and important to attempt to determine what the matching trace evidence means. What is its significance? What is the likelihood that the matching evidence could have occurred by chance and did not originate from the object to which it was matched? Evidential value will depend on the specific hairs and fibers recovered and matched in a case and the circumstances, all of which can vary tremendously from case to case.

HUMAN HAIR

What is the evidential value of a hair match made by comparing microscopic characteristics? What are the factors involved in assessing the evidential value of forensic hair evidence? Full-length hairs, especially if they are long, can often produce strong evidence of association. Because of their length there are more characteristics, and they may vary considerably, along the length of the hair. This variation gives the examiner many more characteristics to compare. Hairs which have been chemically treated by dyeing, bleaching, streaking, etc., may have natural characteristics as well as characteristics due to the treatment. Unusual features, either in the natural characteristics or environmental factors, give hairs greater evidential value because the chance of a coincidental match is less likely when dealing with features that are rare. Experts may disagree about what is unusual or uncommon and the answer often lies in training and experience. Uncommon features, however, would include those features which are easily recognized but which are not regularly seen in hair samples.

Damaged hairs may be considered unusual and may result in greater evidential value when involved in a hair match. Damaged hairs can also provide information that may relate to what happened during the criminal activity, especially in a violent crime. Damage that appears in the known sample and in the questioned hair(s) is significant because a damaged hair is unlikely to occur by chance. Forcibly removed hairs and artificially treated hairs also are uncommon and can result in strong associations. Treatment may also provide information about the time since treatment: Hairs typically grow at a fairly constant rate of ½ in. per month and the untreated "new" portion can be measured and the time frame calculated. This is only an estimate of the time since treatment but it may provide useful information to the investigators.

Gaudette, who has written extensively on forensic hair evidence, has presented various factors that tend to strengthen or weaken conclusions resulting from a hair comparison (Gaudette, 1985). Gaudette uses the phrase "positive hair comparison" to mean that a significant correspondence of hair characteristics exists, not that the hair positively originated from the person with whom it was associated. According to Gaudette, some factors that tend to strengthen positive hair comparison conclusions include:

- Two or more mutually dissimilar hairs found to be similar to hairs in a known sample.
- Hairs with unusual characteristics.

- Hairs found in unexpected places.
- Two-way transfer for example, a victim's hair found on an accuser's clothing and accuser's hair found on the victim's clothing.
- Additional examinations such as DNA testing.

Some factors which tend to weaken positive hair comparison conclusions:

- The presence of incomplete hairs.
- Questioned hairs that are common featureless hairs.
- Hair of non-Caucasian racial origin.
- A questioned hair found in conjunction with other unassociated hairs.
- Known samples with large intra-sample variation.

Some factors which tend to strengthen normal, negative hair comparison conclusions:

- Known sample has more than the recommended number of hairs.
- Known sample shows little intra-sample variation.
- A questioned hair has macroscopic and microscopic characteristics very dissimilar to those of the known sample (length, treatment, other types of analysis shows differences).
- Two or more questioned hairs found together in a clump are dissimilar to the known sample.

Some factors which tend to weaken normal, negative hair comparison conclusions:

- Deficiencies in the known sample: not enough hairs, hairs not representative, hairs are incomplete, a large time difference between offense and procurement of known sample.
- Incomplete questioned hairs.
- Questioned hair has macroscopic and microscopic characteristics close to those of the known sample.

If a hair examiner finds that a hair match is the basis of a strong association, what does the hair examiner rely on to support this conclusion? Their experience as a trained hair examiner and success in numerous test situations are the principle factors. One aspect of this experience is obtained by regularly comparing known hair samples (head and pubic hairs) from different people and finding hairs in one sample to not match hairs in a second sample. The first part of a hair comparison would be to examine all the known samples being submitted to make sure that hairs from one of the samples are not similar to hairs in any of the other samples. Hair examiners regularly find these differences in their everyday examination of hairs from suspects and victims. If head hairs and pubic hairs from different individuals did not generally differ as much as they do, an examiner would see many more instances where single hairs from one person would match hairs in a known sample from another person.

Experience is also important in determining features that are uncommon and which can result in more meaningful evidence. One can also rely on the publications of other experienced hair examiners such as Gaudette and Keeping (1974), and Bisbing (1982) that discuss studies and exercises that have been performed. These studies could not have been accomplished without tremendous variety being present in the population of human head hairs and the ability of trained hair examiners to be able to routinely differentiate hairs from different individuals. It is difficult, however, to know whether a particular examiner has the required level of training, experience, and skill to accomplish the same results as these authors.

One needs to do more than rely on experience alone, however. As Evett has pointed out: "For an expert to say, I think this is true because I have been doing this job for x years is not scientific." He adds, "on the other hand, for an expert to say I think this is true and my judgment has been tested in controlled experiments is fundamentally scientific." A second aspect of experience, therefore, is the knowledge and confidence gained by being tested regularly with human hair matching tests and learning from them. This is most important: It is the only way an examiner can determine what differences are significant and what differences are not, it is the only way to gain confidence in one's ability to get the right answer. It is extremely important that hair examiners be tested extensively during and after their training so laboratory management can determine when an examiner is ready to conduct comparisons and assess an examiner's proficiency. Performance is the key in forensic hair comparisons and the only way to measure performance is with regular testing. And testing of hair examiners requires the development of meaningful proficiency tests.

Proficiency tests are presently required in other forensic sciences, such as DNA analysis, and are being produced and sold to laboratories on a regular basis by outside private vendors. This is not the case with proficiency tests for hair examiners, and it is much more important for the hair examiner to be regularly tested than the DNA expert. Hair comparisons are much more difficult than the comparison and interpretation of DNA typing results. There are difficulties in setting up human hair proficiency tests to be given to a large number of individuals. It is difficult to set up proficiency tests that are equivalent; the variation that may exist in a known hair sample can be difficult to "mass produce". The unknown hairs in each test are going to be different. Some questioned hairs will result in better (and easier) matches than other hairs and the tests, therefore, will not be equivalent. It is easy, however, to set up proficiency tests *within* a laboratory. All that is required are several known head or pubic hair samples consisting of 50–100 hairs and many single hairs from different individuals. With these samples numerous tests can be set up which closely mimic actual casework. The same test can be provided to different examiners within a laboratory to evaluate the test, as well as the examiners. Tests can also be shared between laboratories. Although these tests will still be criticized because the examiners know they are being tested and because they are in-house tests, they still are the best way for the laboratory to determine an examiner's ability. Laboratory management must be responsible for the skills and performance of their analysts and proficiency testing of hair examiners is the best way to measure performance.

TEXTILE FIBERS

There are many factors that influence the evidential value of fiber evidence. Most experienced forensic fiber examiners believe that textile fibers, because of the tremendous variability that exists in the textile industry, can result in strong evidence of association between individuals and objects involved in criminal activity. However, like all types of evidence, the significance of matching fiber evidence can vary to a considerable degree. After a match or matches are made in a forensic fiber comparison using discriminating procedures, it is necessary to attempt to determine the evidential value of the fiber evidence or what might be called the strength of association made through the fiber evidence. One must also be able to provide evidence to support the assessment of evidential value. Quite often a discussion of evidential value is omitted completely in the reporting of laboratory results and in the testimony of those results. If an opinion is presented about the evidential value of fiber evidence in a trial it is often presented with no supporting evidence: It is important that the expert provides the significance of the evidence and not leave to the jury to guess at what the evidence means.

This is one of, if not the most, important part of the analysis of any type of associative evidence. But it can be difficult to determine the evidential value and provide a basis for one's conclusions and interpretation. Quite often one uses experience as a basis for conclusions and often the use of experience is justified. However, how is one to evaluate an analyst's experience? Fortunately, research that deals with the evidential value of fiber evidence has been part of the forensic science literature for many years and especially in the last ten years has generated considerable data for discussion and debate but the topic was also addressed many years ago.

In 1965, Frei-Sulzer wrote a paper in which he addressed many aspects of forensic fiber analysis (Frei-Sulzer, 1965). He pointed out that the evidential value of fiber evidence depends on:

- the scientific comparison of color,
- the rarity of the fibers,
- the location of fibers,
- the number of different fiber types involved in an association.

There have been many other forensic scientists who have addressed the issue of evidential value. For example, Stoney made the following points in considering evidential value in general (Aitken and Stoney, 1991).

- Interpretation of associative evidence is the forensic scientist's greatest responsibility.
- Examination reveals a correspondence in properties, but are these properties common or rare?
- How likely is it that such a correspondence could have occurred by chance?
- An intimate knowledge of the evidentiary material is needed to answer these questions.
- Trier of fact must rely almost entirely on the scientist.
- Interpretation should be as objective as possible.

It would be extremely useful if one could obtain frequency of occurrence data for fibers involved in a specific case. Attempts have been made to create databases that offer estimates of frequencies of fiber types (Biermann and Grieve, 1996a and b, 1998). Some believe that average frequencies should be assigned to fibers and this is probably the one of the best approaches in evaluating the significance of fiber evidence. Fibers would be placed in classes and the evidential value would depend on the class in which a fiber is placed. Although there is little specific frequency of occurrence data available, there have been a number of studies reported over the last 15 years that serve the same purpose. In a sense these studies can be substituted for frequency of occurrence studies at least to assign an average frequency of occurrence to fiber types that are not common types. These studies are called target fiber studies and have demonstrated that there is only a small probability of finding a particular fiber type by chance (Houck, 2003). Target fiber studies involve fiber types that would not be considered uncommon. Most of these studies have occurred in Europe but because there is probably more variety in textile fibers in the US than in Europe, and these studies should also be applicable to the US.

There have been enough target fiber studies to support a general principle of forensic fiber analysis. As long as the fiber type involved in a fiber match is not common, the fiber match can be the basis of a strong association. In other words, it is unlikely that the fiber type involved in the match would be found in a particular location by chance. An experienced fiber examiner examines hundreds if not thousands of textile materials in a typical year of working cases and rarely sees the same fiber type that was involved in an association in one case and in another case.

The more uncommon the fiber, the more likely the association, based on that fiber, is not a coincidence and the greater the evidential value. If a fiber can be shown to be uncommon, therefore, it will have greater evidential value. An examiner should have a firm basis for determining that a fiber is uncommon. There are several factors to consider in classifying a fiber as uncommon. Fibers that are expensive, very old, or those that have not been manufactured for a long period of time are uncommon fibers. Data of this type that supports the rarity of a fiber is not readily available and usually requires that the information be obtained from sources in the textile industry.

Using a classification system, where fibers are classified into general groups such as common, not common and rare, can be used to assess evidential value. Common fiber types are those that are used in the composition of many objects and would be found in the debris from most objects. Fiber types in this category would include white cotton, off-white cotton, certain blue cotton fibers (typically dyed with natural or synthetic indigo dyes), and some undyed synthetic fibers. These fibers would have limited value for association purposes in most situations; that does not mean, however, that those types of fibers should not be searched for and recovered. If one would expect a transfer and retention of these common fiber types under a particular hypothesis, then certainly one should look to see if they are present. In addition, some variability exists within this category, such as, although many blue denim fibers are apparently dyed in the same way, there are some blue cotton fibers that are dyed differently and would have greater evidential value (see Houck's Chapter in *Mute Witnesses*, 2000, for example).

Uncommon fiber types are the opposite of the common fiber category: Although they are "common," in the sense that they occur in many, if not most, textiles, they are easily distinguished from each other. An example of a natural fiber in this category would be cashmere, hair from the Angora goat, which is uncommon because of its high cost. It is not used to any great extent in the manufacture of clothing or other textile items. There are other uncommon fiber types that may be significant because they are not normally seen in debris from clothing. Even though these fiber types may have been used in the construction of many items, their presence on an object would tend to place that object in a particular environment. Trunk liner fibers, which will be discussed later as a part of this case, would fall into this category. Trunk liner fibers usually have features which can be used to identify them as originating from an automobile trunk: They are typically very dark, heavily pigmented fibers or mixtures of black and white fibers and are not generally used for any other purpose. If trunk liner fiber types are unique to automotive trunks, and most are, their presence on a body essentially places that body in an automotive trunk and the association would be recent. Trunk liners are often damaged and therefore trunk liner fibers are readily transferred. Until recently, there was considerable variability among fibers used in trunk liners; in fact, some trunk liner fibers used in the 1970s and 1980s that have been discontinued, would be extremely rare. Evidence to support such claims has been obtained by establishing a database of trunk liner fibers.

Fibers that would fall into the rare category would be fiber types that have not been manufactured in large amounts or which have not been manufactured in recent years. For example, the Monsanto Chemical Company discontinued a carpet fiber with a triskelion cross sectional shape in 1985. Only Monsanto manufactured this particular cross-section. Because of its age and because carpets have an average life span much less than 15 years, there would be very few carpets currently in existence with this cross section. Matching one of these fiber types in a case would be extremely strong evidence of association. There are many fiber types that are no longer being manufactured and would have a small chance of being found in a randomly selected environment.

The most important category of fibers is the colored manufactured fiber, which includes fibers other than those in the common category but for which there is no evidence to support a rare classification. This category would include manufactured fibers such as polyester, nylon, or acrylic that have been printed or dyed. Undyed fibers are made in large amounts but when the fibers are dyed, the resulting fiber types have much greater value for forensic purposes; the total production run for any particular dye may not exceed 8–10 tons (Aspland, 1981). The presence of color makes these fibers valuable for association purposes. There are many different companies both inside and outside the United States that dye or color fibers, textiles, or yarns. Dyeing methods vary not only from company to company but from batch to batch (Aspland, 1981). For example, if 20 companies were to produce a type of garment using the exact same manufactured fiber and attempt to produce a textile material dyed to the same shade, the companies would almost certainly obtain different results (Connelly, 1997). If discriminating procedures are used to compare color, these otherwise similarly colored fibers could be easily distinguished from each other. Polymer types, such as polyester, may be seen often in the composition of clothing and other textile materials but it is difficult to get specific production information about them. Any colored manufactured fiber type is an extremely small percentage of all the fiber types that exist. The probability, therefore, of selecting a person wearing a garment with a particular colored manufactured fiber type or owning a home or vehicle that contains a particular manufactured fiber type would be very small. This is especially true given the fact that fibers do not usually accumulate and persist on clothing for long, textile materials become worn, and may be discarded after several years. The tremendous variety that exists in the population of all fibers should be evident to most people: it is partly this variety in textiles that serves as the basis for our textile retail choices. Juries understand this and asking them to recall how often they have met someone wearing the same item of clothing or someone who has the same residential carpet or owns the same type of automobile easily makes the point. These types of coincidental "matches" occur but not very often; even if two garments appear to be the same, a detailed microscopic examination would typically show otherwise.

The target fiber studies mentioned earlier support the above conclusions about the rarity of colored fibers. Cook and Wilson published details about such a study in 1986 (Cook and Wilson, 1986). This was one of the first studies to evaluate the likelihood of finding particular fiber types by chance. The authors selected four textile items. The items were woolen and acrylic sweaters that would be expected to shed fibers easily and had large sales in England. They characterized the fiber types in the composition of these garments and then searched debris that had been collected from 335 garments that had been received in casework in their laboratory. Of the five fiber types present in the four sweaters, only seven fibers were found that were matching to one of the sweaters. Two red acrylic fibers and another wool fiber matching two of the others sweaters were also found. Fibers matching more than one target garment were found on only one item and there was no more than two fibers of any one type on the items searched in the laboratories. This indicated to Cook and Wilson that to find any more than a small number of matching fibers by pure coincidence is very unlikely. They concluded that when large numbers of fibers matching an item of clothing are found on tapings from a garment in casework, it appears that this is strong evidence of contact. The evidence is greatly enhanced if fibers from more than one item of clothing are the basis of the association. When a two-way transfer of fibers has been established, therefore, there can be little doubt that contact has occurred. The Wilson and Cook study was followed by many similar studies and the results were also similar. These additional target fiber studies included the examination of debris from car seats, cinema seats, outdoor surfaces, pubs and the soles of shoes.

Cotton is the most common fiber type seen in the forensic laboratory and is much more common than man-made fibers (Roux and Margot, 1997). With cotton fibers, the primary characteristic that is being compared is the color. Studies have shown that in comparing these fibers a discriminating way of comparing color is needed to supplement comparison microscopy. If such a system is used, these fibers also have value again because there is still considerable variation even within the cotton classification.

In addition to the commonness of the fiber types, evidential value is affected by other factors. In 1975 Pounds and Smalldon published a paper dealing with the transference and persistence of textile fibers (Pounds and Smalldon, 1975). They found that the initial rate at which transferred fiber are lost during wear is rapid and that in most situations 80% of transferred fibers lost within 2 h. In their study the largest amount of persistence after 34 h was 3%. This was the first of many studies that have been published that deal with the transference and persistence of textile fibers. It is important, therefore, to consider the time and circumstances after transference would have taken place if contact had occurred. Several general principles can be derived from these studies: the fibrous debris on a person's clothing is constantly changing so that an association based on fiber evidence should be considered a recent association. Fibers on a person's clothing reflect recent environments and a victim of a homicide should have fibrous debris picked up shortly before, during, or after the victim was killed (see Deedrick's Chapter in *Mute Witnesses* for additional principles on trace evidence transfer).

THE CASE

The case occurred in South Dakota in 1985. The following material is taken from several legal decisions about this case. On the morning of May 8, 1985, the victim left his home to attend classes at a nearby college. He never returned. After finding her son's locked car near their home his mother contacted the Sheriff's Department. Two days later she was told in a phone call to go to a phone booth, retrieve an envelope and she would find out what had happened to her son. The victim's mother found a ransom note demanding $200,000 in exchange for her son; $190,000 was obtained from local banks and delivered to the mother's residence. The mother received a second telephone call 3 days later in which she was instructed to leave the money near a statue in a park, telling her what her son was wearing but refusing to let her speak to her son until he got the money. The money was left in the requested location along with a radio-tracking device under a false bottom in the briefcase. An individual in a 1975 blue Oldsmobile picked it up and was arrested later that day. The arrested subject was found with $760 of ransom money in his wallet and most of the rest of the money in his car.

The victim was found on May 28, 1985, when a farmer found his bludgeoned, maggot-ridden corpse in an old icehouse. He had been shot three times and had head wounds inflicted by blows from a square edged instrument. There was no indication that the victim had been killed at the icehouse – the body apparently had been dumped. The corpse was in a partially mummified state and showed no signs of having been tied up. Stomach contents included part of an olive, degenerated vegetable matter, and meat fibers, which were consistent with the victim's last known meal: A pizza topped with pepperoni, sausage, mushrooms, and olives consumed on May 8.

Evidence submitted to the FBI Laboratory included the following items:

- hair samples from the subject,
- hair samples from the victim,
- sample from a throw rug from the Oldsmobile trunk,
- a trunk liner sample from 1975 blue Oldsmobile,
- the victim's clothing (shirt and blue jeans),
- hairs from various locations (from trunk and door frame used to cover the body),
- fibers and a thread from the door covering the victim,
- purple gloves from the subject's Camaro,
- vacuumings from the Oldsmobile trunk,
- debris from the Oldsmobile trunk liner,
- hair and fibers from the Oldsmobile trunk liner.

The South Dakota Forensic Laboratory had begun to process the hair and fiber evidence and had mounted some of the samples. The mounting of hairs and fibers is accomplished by placing fibrous materials in a mounting medium between a glass microscope slide and a glass cover slip. It is necessary to use a mounting medium with a refractive index near 1.52 to be able to obtain good images and to examine and compare the microscopic characteristics and optical properties exhibited by fibers. Samples of fibers from various items from the victim's home were also submitted for elimination purposes. These items included: carpet samples from a residence that the victim was at prior to his disappearance, carpet and upholstery samples, and trunk liner fibers from his mother's 1985 Oldsmobile.

LABORATORY APPROACH

Although most laboratories use similar procedures and there have been attempts to standardize the fiber analysis methods (SWGMAT, 1999), there is still considerable variability in how different forensic scientists and different laboratories approach a case. This is primarily because training and, more importantly, the knowledge and experience of hair and fiber examiners are always going to be somewhat different. Most believe that it is important to become familiar with the circumstances surrounding the case prior to beginning the analysis. Given a scenario provided by investigators to the laboratory, one would look for hairs and fibers that would be consistent with contact between the victim and the subject or with items associated with the victim or suspect. Depending on the circumstances, if one does not find transfer evidence supporting a particular scenario, it may be necessary to modify or change one's hypothesis about the crime. In some cases, the absence of trace evidence may be sufficient to eliminate a person as a suspect. Of course, it should also be evident that, because trace evidence can be lost easily and quickly, the absence of trace evidence on an object does not necessarily mean there was no contact with another item.

One also needs to consider if other evidence might be important in the case. If there is other evidence in a case such as DNA or fingerprints and identification is the primary issue, it may not be necessary to spend a large amount of time on the trace analysis. Because of the power of other types of evidence for identification purposes and depending on the circumstances and issues to be resolved in a case, trace evidence may provide little added value. The need for trace analysis examinations often depends on what questions are being asked about the crime. It is also important to become familiar with the types of objects involved and their transference and persistence characteristics. One needs to consider whether the trace evidence provides any information on the timing or sequence of different events that may have taken place during or after the commission of the crime. One may also need to consider whether the evidence collection procedures at the crime scene and at the Medical Examiner's Office were adequate.

Other factors that must be considered include the likelihood of contamination occurring sometime before the evidence was received in the laboratory. Could contamination have accounted for the results in the case? Were procedures designed to prevent contamination in place? What was the condition of objects involved? Were they damaged and/or old such that one would expect a large amount of transfer? Are additional searches or recovery efforts necessary? Are the known textile samples suitable if the entire item is not submitted?

One must also decide on the recovery and handling procedures to use in the laboratory. The procedures depend, primarily, on the type of evidence being processed, the condition of the evidence and the potential for contamination in the laboratory. The order in which the evidence is being processed must be determined, again keeping in mind the potential for contamination. Whatever procedure is used to collect the trace evidence, the mounting of a large representative number of debris fibers has many advantages. The actual number of fibers mounted would depend on the amount of debris that has been recovered. It is quite easy to mount large numbers of fibers if a scraping procedure is used to collect the debris. But scraping can only be used if proper facilities are available and scraping is only suitable for certain types of objects. Large numbers of fibers can also be mounted when one is examining small objects under low magnification and fibers removed with tweezers can be easily mounted on glass microscope slides.

There are many objects, however, where an efficient recovery of trace evidence requires a taping procedure to be used. Fibers can be easily removed with tape and taping is very efficient. However, considerable time is often required to remove the fibers from the tape and mount them on a slide. Most often a representative sample of fibers is mounted but if all of the debris can fit on one or two slides it should all be mounted. Screening of the fibers on a slide would then occur at a magnification between $40\times$ and $100\times$ natural size. If a large number of fibers are mounted, one is less likely to miss important fibers and most importantly one obtains a complete picture of what is present in the debris. This is something that cannot easily be done without mounting large numbers of fibers. In addition, when all hairs and fibers are embedded in a semi-permanent mounting medium there is little sample handling and the fibrous materials cannot be lost. The handling of trace materials requires considerable time and increases the chance of loss. Most of the microscopical comparison procedures can be conducted on the mounted debris, which can be easily removed from most mounting media if additional testing is required.

Usually known materials should be mounted after the questioned materials. If precautions are not taken and proper procedures not followed, fibrous materials may be lost during the mounting process and could possibly contaminate the debris from questioned items if the known materials are handled first. If one is interested in identifying colors of interest during the searching process, this can be done when descriptive notes are being taken during evidence recovery.

Taping usually results in fewer debris fibers being mounted because of the time involved in the recovery of the fibers from the tape; however, often the best and most efficient way to recover fibers is with tape. With taping, the number of fibers mounted depends on the circumstances of the case. It is however, very efficient and less prone to contamination and loss than other procedures.

Screening is accomplished with a microscope and the mounted fibers are examined using magnifications of $40-100\times$ normal size. Even if hundreds of fibers are mounted, they can be examined in a short period of time. It is usually not necessary to do any characterization during this screening process although a good microscopist can easily identify most fibers microscopically. The process is easier if a comparison microscope is available for the screening as comparisons can be conducted at the same time as the screening. Fibers of interest are marked, including frequently occurring fibers that are foreign to the object from which the debris was taken. One must also decide whether additional debris slides need to be prepared for analysis; usually this is done when there is some indication of an apparent transfer but only a small number of matching fibers have been recovered. Using the comparison microscope one then compares the microscopical characteristics and optical properties of any similar questioned and known fibers.

Optical properties of any matching fibers can then be examined and compared with a polarized light microscope. The optical properties allow for the identification of most manufactured fibers and are also important for comparison purposes. Although the comparison microscope is very discriminating in the comparison of the color of manufactured fibers it is not as good with colored natural fibers. A microspectrophotometer, which is attached to a microscope, is an easy and very discriminating procedure for the comparison of color and as indicated, earlier color is the feature that gives fiber evidence great variety and value (Siegel and Houck, 2001). Both questioned absorption curves and known absorption curves obtained using the microspectrophotometer should be on same chart paper for easier comparison purposes. A fluorescence microscope can also be used to look at the fluorescence of the known and unknown fibers. Notes should be taken on the characterization of probative fibers and on numbers of recovered matching fibers.

ACTUAL HAIR AND FIBER RESULTS

The first FBI Laboratory Report in the case was issued on February 19, 1986. Matching hairs or fibers in a case either originated from the source that they were matched to or are present by coincidence. In a coincidental match, the matching hairs or fibers happen to be present but they actually originated from another source. As Gaudette has pointed out, in making this statement, two assumptions are being made (Gaudette, 1995). One is assuming that the comparison procedures are very discriminating so that differences are unlikely to be missed and that contamination has not occurred; if proper methods of processing and comparison are used these assumption are very likely to be true. If the alternative to an association is a coincidental match, it becomes important to try and estimate the chance of a coincidental match. The smaller to chance of a coincidental match the more powerful the evidence of association. The chance of a coincidental match depends in part on the frequency of objects that exist with similar fibers. With the tremendous variety that exists in the world of textiles, any one fiber type as long as it is not common, is an extremely small percentage of all the fiber types that exist (Houck, 1999). Therefore, the probability of finding a particular fiber type by chance in the debris from a particular object or in the composition of a particular textile material is going to be small (Roux and Margot, 1997); accordingly, the probability of a coincidental match will also be small. Another point about transferred trace evidence needs to be emphasized again. As shown by several studies, trace materials are easily lost as someone goes about their daily activities. Clothing items are also routinely cleaned which will remove most debris, therefore, the debris on an object is constantly changing.

In discussing the results in this case, the location of hairs and fibers will be indicated in parentheses. Throughout this report fibers and animal hairs are reported as being like other fibers or animal hairs. The criteria used to make a determination that a fiber or animal hair is consistent with having originated from a particular source involve the comparison and matching of numerous microscopic characteristics and/or numerous optical properties, which hair and fibers exhibit.

The FBI report stated as follows:

> Three light brown to brown head hairs of Caucasian origin were present on the Q12 glass microscope slide (trunk vacuuming). One light brown head hair of Caucasian origin was present on the Q15 slide (debris from driver's side of trunk liner). These four hairs have been severed at their proximal ends. These hairs exhibit the same microscopic characteristics as present in the K4 hair sample (known head hair sample from the victim) and, accordingly, could have originated from the source of the K4 sample." It is pointed out that hair comparisons do not constitute a basis for absolute personal identification.

The last statement is a comment that the FBI Laboratory includes in all reports involving a hair match which is added because hair examinations are not a method of positive identification. A match has been reported out between hairs recovered from the trunk and the known sample from the victim. The hairs involved have all been severed at their root ends and are consistent with originating from the victim. There was no discussion of the significance of these results in the report. There was no information provided in the report about the likelihood of these hairs originating from someone other than the victim. The significance of the hair match was not addressed in my report other than pointing out that the hair match is not an absolute identification. At the time this report was written, it was not common to discuss the significance of one's finding in the laboratory report. The meaning of matching evidence is still not addressed in forensic laboratory reports dealing with many types of trace evidence. It is this author's opinion that some indication of the significance of the matching evidence should be reported.

The significance of a hair match falls somewhere between having very little significance (a very common item) and having very high significance, an identification (a unique item). This author believed that the four matching severed head hairs, as set forth above in this case, were strong evidence that the hairs originated from the victim and that was how the author testified at the trial. It was the author's opinion that the hair evidence alone was the basis of a strong association between the victim's body and the automobile's trunk, where they were recovered. In other words, these results were not expected to occur by accident.

This author's reasoning, which was presented at the trial, was based on several factors. The probability of finding hairs matching the victim in a randomly selected trunk would be small. The fact that four hairs were found increased the evidential value as each of these hairs all matched the victim's known sample. Contamination, at least from the known head hair sample of the victim, which hairs had roots, could not have occurred. The fact that the hairs are severed is also important: It is exactly what you would expect if the victim with a serious head injury had been placed in the trunk. As a coincidental finding, it is very unusual to find Caucasian head hairs without roots in debris from clothing or some object. Caucasian head hairs do not fragment easily; hair clippings, like those remaining after a hair cut, are sometimes recovered but they are usually much shorter, found in larger numbers, and usually severed in a similar manner on both ends. The hair evidence in this case is exactly what you would expect to find if someone with a head injury were placed in the trunk of the Oldsmobile. This same evidence would be unlikely to be found in a randomly selected car trunk.

A defense expert reviewed the FBI Laboratory Report and included his assessment of the significance of the finding along with results. He reported that on a slide containing debris from the trunk, "... *four of the human hairs were similar to the head hair standards of the victim.*" His report also provided his assessment of the hair evidence as follows.

The finding of hair similar to the victim's in the Oldsmobile trunk must be interpreted with caution. On one hand, the hairs that were found were all severed at the proximal end. Such severing is consistent with, but not proof of, a blow to the head with a hard object. Since human hairs cannot be identified as coming from a specific individual, it is incorrect to conclude that the hairs are the victims. Based on the observable microscopic features in these hairs, they could be from the victim or anyone with similar hair. The percentage of the population with similar hair cannot be stated accurately, but is probably in the range of 5–20% of the Caucasian population.

There is agreement that hairs matching the victim were found in the trunk but the defense argued that these hairs are very common. If these hairs, or for that matter any Caucasian head hairs, were that common, then hair would have very limited value in making associations; this is not what has been subsequently reported (Houck and Budowle, 2001). The defense expert did not testify in the case; however, one way to counter this type of testimony is to point out to the jury that by just looking at gross characteristics, such as length, color, and curl, one can classify hair in more categories than claimed in the defense expert's report. And when a hair comparison is conducted, the examiner looks at many more characteristics other than length, color, and curl. It would also be impossible for trained hair examiners to perform as they have in proficiency tests and in the reported literature if hairs varied as little as claimed by the defense expert. As mentioned earlier, an experienced examiner typically finds that head hairs from different individuals exhibit significant differences. Given the severed ends and the head injury to the victim, the association became much stronger. If this is true, then why, one may ask, are there so many questions about the reliability of forensic hair examinations? This author maintains that the simple answer is that there are many poorly trained hair examiners that are lacking in ability in the field.

The author's report continued:

> Fourteen black rayon fibers were found on the Q3 and Q4 glass microscope slides and in debris from Q3 and Q4 (*Q3 and Q4 are the blue jeans and shirt recovered from the victim's body*). These rayon fibers are like black rayon fibers present in the composition of the K5 trunk liner and, accordingly, could have originated from the source of K5.

Black rayon fibers found on the victim's clothes were matched to fibers in the composition of what was left of a trunk liner from the Oldsmobile. The trunk liner was made of large diameter, heavily pigmented viscose rayon fibers. Fibers like these are typically not seen in a forensic laboratory: this fiber type was seen only once before by this author in a case involving a trunk liner from a 1977 Cadillac. During the case involving the 1977 Cadillac, junkyard studies had been conducted to try and determine if black rayon trunk liner fibers were common. Additional junkyard studies of trunk liners found in both old and new cars were conducted for this case. It was determined that General Motors used a trunk liner composed primarily of black rayon fibers in many of their vehicles manufactured prior to 1978. But even though this fiber type would have been used in perhaps hundreds of thousand of vehicles, except for the Cadillac, it had not been seen by the author previously in casework. A fiber examiner obviously cannot remember every fiber examined but one can often recall fibers that are unusual. This type of fiber was uncommon and would have been remembered if it had appeared previously in a case.

A large diameter fiber is most often going to be from a carpet but rayon is not typically used in residential carpet and black is not a common color for residential carpet. General Motors used rayon in automotive carpet up until 1974, but those automotive carpet fibers were dyed and not pigmented in the same way as the trunk liner fibers. Over the years I have collected many trunk liner fibers and found that with a few exceptions, black rayon fibers were not used in trunk liners after the 1977 year model. Even though this crime occurred in 1985, 7 years after General Motors last used this type of trunk liner in their automobiles, and approximately 30% of cars originally containing such fibers would have been scrapped, there would still be large numbers of cars with a similar trunk liner in existence. However, when all vehicles are considered, the relative frequency of automobiles with a similar trunk liner would be quite small. Therefore, the chance of randomly selecting a vehicle in 1985 and finding it to have a trunk liner with the same fiber composition as the subject's 1975 Oldsmobile would be small. Given the hair and fiber evidence thus far, there was very strong evidence that the victim was in the trunk of the subject's 1975 Oldsmobile.

The report continued:

> Thirteen yellow acrylic fibers are present in the Q12 debris (from the trunk). Two yellow acrylic fibers are present, one each in the Q3 (victim's blue jeans) debris and in the Q4 debris (victim's blue jeans). The yellow acrylic fibers from Q3 and Q4 are like the yellow acrylic fibers in the Q12 debris and, accordingly, these fibers from Q3, Q4 and Q12 could have originated from the same source.
>
> Three maroon polypropylene fibers are present in the Q12 debris. One maroon polypropylene fiber was also removed from Q4. The maroon fiber from Q4 is like the maroon fibers present in the Q12 debris. Accordingly, these fibers from Q4 and Q12 could have originated from the same source. Possible original sources of the yellow acrylic fibers and maroon polypropylene fibers discussed above are unknown to the laboratory.
>
> Numerous light brown to dark brown animal hairs are present in the Q12 debris. One light brown to dark brown animal hair like those in the Q12 debris was found in the debris from Q3. These animal hairs from Q3 and Q12 could have originated from the same source.

The defense expert agreed that the black rayon fibers recovered from the body matched fibers from the trunk liner and also agreed that matching yellow acrylic, maroon polypropylene, and some animal hairs were recovered from both the trunk and the victim's clothing.

Regarding these fiber matches, the defense expert reported the following:

> In addition to similar fibers which were found on items associated with the victim and in debris from the trunk of the Oldsmobile, a large number of fibers were found among the debris from the Oldsmobile, and on the clothing of the victim, for which no counterparts were noted on the other evidence. In assessing the significance of this evidence, particularly the animal hair, maroon olefin and yellow acrylic fibers, it is appropriate to consider two questions:
>
> ■   Why are there so many different fibers present in the debris from the trunk, which have no counterparts in the debris from the clothing (and conversely)?

- How many vehicles with similar black rayon trunk liners, which might have been used to transport the victim, have in their trunk 2 or 3 different types of hairs or fibers, which would be similar to hairs, or fibers found on the clothing of the victim? The answer to neither of these questions is available.

He is correct in that one cannot produce or determine an exact answer to his questions. This type of criticism of hair and fiber evidence is commonly heard in the courtroom. The defense expert's claim is true that we do not have exact numbers or even good estimates about the frequency of fibers like those in this case in the general population of textile items or in debris. But, again because of the tremendous variety in the fiber world, a specific type of polypropylene or acrylic fiber is not going to be found very often by chance in a particular environment (Roux and Margot, 1997). These researchers and other target fibers studies have shown that a man-made fiber of a particular type and color has a frequency much smaller than 1% of the fiber population and this is before the many comparison features of fibers are considered. The fibers on the victim's clothing help define the environment the body was in shortly before the victim stopped moving around and the combination of the three fiber types in debris from someone's clothing would be found in only a few environments. This has been the conclusion in all target fiber studies.

The other criticism made by the defense is that debris fibers are present in the trunk and on the body that could not be matched to anything. There will always be debris fibers that will be present on or in an object (a body and a trunk in this case) that do not match any of the known fibers, or even the other questioned fibers, in any one case (Gaudette, 1993). In this case some debris fibers are common types that would have little evidentiary value and would not even be compared in most instances; some will be from the victim's environment (residences, workplaces, automobiles, etc.). Depending on the debris fibers, it may be important to obtain elimination samples to try and account for certain foreign fibers. However, it would be almost impossible to account for every fiber type found in a particular environment (Gaudette, 1993); additionally, some of the questioned fibers may be from other items that are related to the crime but which have not been recovered.

In this particular case, if the victim's body was placed in the field shortly after May 8, the body would have been exposed to the weather for 20 days, it may have been disturbed by animals, and some evidence could have been lost from the victim's clothing. More important in this case, however, is the likely change in the fibrous environment of the Oldsmobile trunk. There was a cursory search of the trunk of the Oldsmobile on May 14, 1985, but nothing was taken at that time. During a second search on June 18, 2000, it was observed that the Oldsmobile's trunk carpet had been removed and replaced with a throw rug, the tire iron was missing, and a white powder was sprinkled in the trunk. There was testimony at the trial that the trunk carpet was removed and baking soda (identified by the FBI Laboratory) was added to the trunk to remove a bad smell. The debris, submitted to the FBI Laboratory, was not recovered from the trunk until it was vacuumed on July 29, 1985. The contents of the trunk had certainly been modified in the 2-month period since the prosecution claimed the trunk was used to carry the body and, given the circumstances, it is likely that the debris recovered from the trunk was only a small part of all the debris that was present earlier.

Blood samples were also recovered from the trunk on July 15, 1985. This blood turned out to be unsuitable for the genetic marker testing that was a part of a forensic laboratory testing protocols at the time. If the current methods of DNA testing had been available in 1985, it is likely that DNA typing results matching the victim would have been obtained and the hair and fiber evidence would have had less significance but still would have been important.

The defense expert continued:

> Similar animal hair and three similar types of synthetic fibers were found both on the victim's clothing and in the trunk of the automobile. Hair similar to the victim was found in the trunk of the car. These findings suggest the possibility that, sometime prior to his body being left at the spot from which it was recovered, the victim was either in, or in close contact with something else from, the trunk of the car. However, especially given the known prior association of the victim and the subject these finding must be interpreted with caution.
>
> All of the items found could remain attached to the clothing for a relatively long time. None of the items are unique, and for some the source is entirely unknown. Even assuming that the victim was transported to the place where his body was found in the trunk of a vehicle (a reasonable assumption based on the presence of fairly large number of black rayon fibers on his clothing), no information is available which would allow an estimate of the likelihood that other vehicles with the same type of trunk liner would also have fibers which are similar in color and type to fibers found on the victim's clothing.

The defense points out that all of the findings in this case could have occurred by chance and claims that there is not enough information available to estimate the likelihood that similar findings could not be obtained with other vehicles. In other words, the defense claimed it is possible that the hair and fiber evidence could have had nothing to do with the subject. In a situation like this it is necessary to point out that it is not meaningful to talk about *possibilities*: one must address *probabilities*. What is the probability that the findings in a case are coincidental and have nothing to do with the subject? Even though we cannot come up with an exact probability, each of the findings in this case would be unlikely to occur by chance. The defense conclusion also fails to address that the likelihood of obtaining *all* of the hair and fiber matches in this case by chance is much smaller than any of the individual findings. Even if numerical estimates of probabilities cannot be generated, that does mean one cannot use reasonable qualitative (non-numeric) estimates.

If one were to conservatively assign a frequency of 1/100 to finding severed head hairs like those in the suspect's trunk and assign a frequency of 1/100 to finding each of the three manufactured fiber types linking the victim to the trunk and then assume that each of these findings are independent (that the fibers all came from different, otherwise unrelated textiles), the product rule would give a frequency of 1 in 100 million of all of these events occurring at one time. While these assumptions are not based on any hard evidence, and the frequency of each of the evidence types is likely to be much smaller, it does give the reader an idea of the relative frequency of combined evidence.

One additional finding was reported by the author:

> The Q3 pair of blue jeans is composed in part, of blue cotton fibers. Numerous blue cotton fibers like those in the Q3 blue jeans were present in the Q12 debris and, accordingly, could have originated from Q3. It is pointed out that blue cotton fibers like those in the composition of Q3 are common and are only of limited value for association purposes.

This is a finding that one would expect if the victim's body had been in the trunk. However, these blue cotton fibers are common and of limited value for association purposes.

Testimony by the author, regarding the hair and fiber evidence at the trial, stated in part that, "it is the basis of an extremely strong association between the trunk of the car and the victim, and in my opinion, the probability or chance that the victim was not in the trunk in this case is extremely small."

The hair and fiber defense expert in this case used the term possible or possibility on several occasions with respect to the findings in this case implying that, while it was possible the association indicated by the hair and fiber evidence was real, other explanations were possible to explain the findings: this is true. One could always come up with a scenario to explain a particular set of findings. But this often begs the question: what is the simplest explanation that fits the circumstances and evidentiary facts in the case?

Many advocate the likelihood ratio (LR) approach to the assessment of evidential value. The advocates of the LR approach, which have been discussed in numerous publications over the last 15 years, believe that a forensic scientist cannot speak about the probability of a particular hypothesis, as was done in this case. They believe that the forensic scientist must only assess the probabilities of obtaining evidence in a case under two (or more) contrasting hypotheses. An LR in this case could be obtained by first calculating or estimating the probability of the hair and fiber evidence under the (prosecutor's) hypothesis that the blue Oldsmobile was used to carry the victim's body. This probability would certainly be large and approach 1 (or 100%). In other words, these findings would almost certainly occur if the victim's body had been placed in the trunk. One must then calculate the probability of obtaining the hair and fiber evidence given a (defense) hypothesis that provides alternative sources for the circumstances. In this case the defense might claim that a different vehicle, one that was not linked to the suspect, was used to transport the victim. The probability of obtaining these results if some randomly selected vehicle was searched would be extremely small. An LR would be calculated by dividing a very large probability by a very small probability to produce a large LR. If an LR such as 1000 were obtained, then one could testify that it is 1000 times more likely to obtain the hair and fiber evidence in this case under the prosecutor's hypothesis than it would be to obtain this evidence under the defense hypothesis. If frequency of occurrence data were available, an LR could be estimated but it would only be an estimate and would be based on assumptions. The estimated probabilities would be for events and particular hair and fiber findings that have occurred only once. If one were to use this approach it would be necessary to use some type of average probability and assume that all of the individual events in this case are independent of each other. What is actually being done when one uses the LR approach is probably the same thought process that a laboratory examiner goes through in deciding on the evidential value to assign to the evidence in a particular case. The proponents of the LR approach do not recommend presenting an actual number to a jury but to assign general statements to levels of LR.

Rudram (1996) and Evett (1987) and others have suggested that numerical estimates of the LR not be given but that some verbal expression is assigned to a particular LR. Rudram suggested that a scale of conclusions could be included in the report as follows:

- A did not come from B,
- it is unlikely that A came from B,
- there was nothing to suggest either that A came from B or that it did not,
- A could have come from B,
- A probably came from B,
- A most probably came from B,
- A came from B.

If something like the above list were used, each examiner would have to decide and prepare examples of evidence that would define each of the responses. For example, one could require three unrelated matching fiber types as the basis of statement number 6 – that A most probably came from B.

THE VERDICT

The jury, after a 6-week jury trial that included considerable probative evidence other than the hair and fiber found the subject guilty of first-degree murder, kidnapping, possession of ransom money and forgery. As indicated earlier, the hair and fiber evidence was only a part of the case against the subject; some of the other evidence will be discussed in the remaining sections of this chapter.

THE APPEALS

The subject appealed his convictions and alleged numerous errors had occurred during the trial. The appeal covered many different areas and he sought initially to suppress evidence acquired through various searches carried out under three different search warrants. These warrants authorized the search of the suspect's ranch and vehicles owned by the suspect's father and wife. A warrant also was obtained to search the suspect's residence and that of his father-in-law for bank checks drawn on the subject's account and any writings contained the victim's name. The suspect alleged that there were numerous errors regarding these searches and the warrants authorizing the searches. The South Dakota Supreme Court went into considerable detail in their decision explaining why the errors alleged by the subject were of no consequence, harmless, or his allegations were erroneous. The subject also appealed the forced production of his handwriting samples but the Supreme Court did not address this issue because of the overwhelming evidence against him on the forgery count.

The subject had made motions prior to his trial for a change of venue, additional preemptory challenges, and an expert to perform a jury survey, but the trial judge denied the motions. The defendant had asserted that prejudicial pretrial publicity made a fair trial impossible and argued that all of the selected jurors knew of the case through media coverage. The subject also challenged certain hypothetical questions made by the State during jury questioning that included hypothetical questions of prospective jurors as to whether they had ever seen a person that they thought they knew, called out his name, and discovered that the person was actually someone else. The subject argued that these questions were a tactic used to undermine one of his witnesses in advance. In all of these issues the Supreme Court found none in the subject's favor.

One of the appeals dealt with the victim's clothing on which probative fiber evidence was recovered. He contended the chain of custody concerning the victim's clothing was inadequate, as the clothing was left exposed and unattended on the laboratory rooftop to dry. The court ruled that the victim's clothing, although initially not analyzed at the time of recovery, was sufficiently accounted for during the period of time prior to the laboratory analysis.

Concerning the period it was exposed, testimony about the possibility of contamination, backed by testing, was presented to the trial court during a pretrial hearing. The court ruled that there was no evidence of tampering and little likelihood of access by others. They also believed that it was improbable that the original items had been tampered with or altered. The court argued that the State must only show a reasonable probability that no tampering or substitution has occurred, and the State need not negate every possibility of tampering or substitution. Mere speculation is insufficient to establish a break in a chain of custody.

The subject also challenged the sufficiency of evidence supporting his convictions on the kidnapping and murder counts, suggesting that there were no facts indicating that crimes were committed in Brown County. The court ruled that these arguments were without merit, and pointed out that the subject simply recited whatever evidence was most favorable to him, while deriding unfavorable evidence and inferences therefrom as speculation. The Court concluded that the evidence in the record was more than adequate to support the jury's verdict. The issue of sufficiency of evidence relates to the hair and fiber evidence, as it is the evidence which places the victim's body in the trunk of the automobile used by the subject to pick up the ransom money. As will be discussed later, another judge disagreed with the South Dakota Supreme Court.

The Supreme Court held on August 3, 1988, that:

- The evidence was sufficient to support the convictions;
- The improperly admitted testimony of the forensic anthropologist did not require reversal in light of other evidence concerning to time of death;
- The defendant was not entitled to change of venue and;
- The trial court's use of aiding and abetting instructions was not reversible error.

The subject subsequently filed new appeal alleging ineffective assistance of counsel, which was denied in a written decision on July 10, 1991. The court in denying the subject's *Habeas* petition, found that the subject's trial counsel did an exemplary job in vigorously representing him.

But the United States District Court thought otherwise and the following headlines appeared in the local newspaper on March 26, 1993.

> Insufficient evidence – Kidnapping; murder conviction overturned. …A federal judge upheld the subject's conviction on forgery and possession of ransom money

The newspaper article provided the following information: A US District Judge overturned the 1986 kidnapping and murder convictions against the subject ruling that prosecutors did not present enough evidence to convict him of murder and kidnapping charges. The judge upheld the subject's convictions on forgery and possession of ransom money. The State Attorney at the time of the decision commented after the judge's decision: "We had enough evidence for 12 jury members to convict him, enough evidence for five state Supreme Court Justices to unanimously uphold the convictions on all charges. Now we have one federal judge who has single-handedly overruled them."

The District Court Judge believed that enough evidence existed to convict the subject on charges of forgery and possession of the ransom money but in his nine-page ruling, he stated that:

- No evidence was presented as to who shot or beat the victim.
- Bullets found in the victim's body could not be traced to any specific gun.
- A 0.22 caliber cartridge found in the suspect's car could not be matched to the bullets used to kill the victim.
- No square-edged instrument matching the imprint on the victim's head was ever found.
- No testimony placed the suspect and the victim after the kidnapping or placed the suspect at the site where the body was found.

The District Court Judge was agreeing with a federal magistrate, who had earlier ruled that insufficient evidence was presented at the subject's trial to convict him of murder and kidnapping, and the Judge's decision agreed with the magistrate's findings. Since two of the victim's charges had not been reversed, he was still in jail but he had nearly served his time for the lesser sentences.

It would appear that, according to the logic of the Magistrate and District Court Judge, if a lone person kills someone with no witnesses and successfully hides the murder weapons, he or she could not be found guilty unless he or she confesses. The Magistrate and District Court Judge either ignored or didn't understand the hair and fiber evidence. The hairs and fibers were extremely strong evidence of recent contact between the victim's body and the trunk of the vehicle driven by the subject to pick up the ransom money.

It was the State's turn to appeal, this time to the United States Court of Appeals of the Eighth Circuit. On July 15, 1994, the Appeals Court reached their decision and held that the evidence supported all the charges on which the subject was convicted and there was no constitutional error in giving aiding and abetting instruction. Concerning the homicide and kidnapping, the Appeals Court ruled that evidence was sufficient to support first-degree murder and aggravated kidnapping convictions under South Dakota law. They presented the following basis for their decision:

- The defendant had contacted victim's mother demanding ransom money in exchange for victim's return.

- The defendant retrieved ransom money from designated location and told police five differing and conflicting stories after his arrest.
- Hair, fiber and blood evidence supported the conclusion that victim's body had been in trunk of automobile that defendant drove when he picked up ransom money.
- In addition, the defendant was observed retrieving ransom money from the designated drop spot driving a blue Oldsmobile. He was later arrested in possession of most of the ransom money.

After his arrest, however, the defendant began to weave a tangled web of explanations. At first, he explained that a man, named Mike, had intended to buy a horse from him and had given him some traveling money. At this point, officers asked the defendant where Mike was without mentioning the victim's last name or that Mike was in any sort of trouble. The subject responded, "I wouldn't hurt the victim. He's my best friend."

Next the subject changed the story by telling officers that an acquaintance had asked him to pick up a package in the park, deliver the package, and in exchange, the acquaintance would buy a horse from the suspect for $45,000. In a third version, the subject said that the acquaintance had instructed him by telephone to call the victim's mother, deliver the envelope to the bar, retrieve the money and then deliver it to a motel in Alexandria.

The subject later admitted that he fabricated the horse purchase as a cover story and asserted that the acquaintance threatened to harm victim or the subject's newborn child if he refused to cooperate.

Later the defendant told a totally new story involving two unknown men who accosted him at a bar on May 7, 1985. According to this story, the men threatened to harm the victim and the subject's baby unless he delivered the envelope to the victim's mother and delivered the ransom money to them in Minnesota.

Finally he offered his fifth and final version of the story, implicating the victim, as the initiator of an extortion scheme against his own mother. He claimed that he was threatened into cooperation, and he admitted that he had written the ransom note and made the calls to the victim's mother as part of the scheme. Other testimony at trial indicated that the subject's father was in severe financial difficulty, that the son wanted to help him because a land payment was delinquent and due on May 14, 1985, the day after he retrieved the ransom money.

As pointed out earlier, this was a case with strong evidence of guilt even without the hair and fiber evidence, but the hair and fiber evidence strongly linked the victim's body to the subject, an important aspect of any homicide case. There have been no court decisions in this case since 1994 and, perhaps, finally, this trace evidence case that went on for over 11 years is over.

CHRONOLOGY

5/8/85	The victim leaves home to travel to attend a college class
5/9/85	The victim's automobile is found abandoned near his house
5/10/85	Call made to the victim's mother requesting a ransom of $200,000 for the safe return of her son
5/13/85	Second call to victim's mother advising her of the location to deliver the ransom money
5/14/85	The subject is observed picking up ransom money driving a 1975 blue Oldsmobile
5/15/85	The subject is arrested
5/13–18/85	During a five-day period after his arrest, the subject gives five different statements attempting to explain how he obtained the ransom money, each somewhat different
5/18/85	Trunk of 1975 Oldsmobile searched "with negative results"
5/28/85	The victim's bludgeoned and maggot-ridden corpse is found by a farmer in an old icehouse. He had been shot three times, and had head wounds that were inflicted by square edged instrument. His body appeared to have been dumped at the body recovery site
6/17/85	A search warrant authorizing search of the subject's father's ranch and vehicles owned by the subject's family is obtained
7/3/85	A second warrant authorized the search of subject's family cars, a Camaro and the Oldsmobile
7/29/85	Trunk of Oldsmobile vacuumed for evidence (but prior to this the "trunk was washed out and trunk liner discarded"
8/2/85	A third warrant authorized a search of subject's residence and that of his father-in-law for bank checks drawn on victim's account and any writings contained the victim's name
1/6/86	Letter from the Office of Attorney General's Forensic Laboratory that accompanied the first submission of evidence to the FBI laboratory
3/18/86	Trial scheduled to begin with an anticipation of approximately two weeks scheduled for jury selection and a list of 70 possible witnesses to be used by the prosecution

REFERENCES

Aitken, L. and Stoney, D. eds. (1991) *The Use of Statistics in Forensic Science*. Chichester, West Sussex, UK: Ellis Horwood.

Aspland, J.R. (1981) "What are dyes? What is dyeing?," in *AATCC Dyeing Primer*. Research Triangle Park, North Carolina, American Association of Textile Chemists and Colorists.

Biermann, T.W. and Grieve, M.C. (1996a) "A computerized data base of mail order garments: a contribution toward estimating the frequency of fibre types found in clothing. Part 1: The system and its operation," *Forensic Science International*, 77, 65–73.

Biermann, T.W. and Grieve, M.C. (1996b) "A computerized data base of mail order garments: a contribution toward estimating the frequency of fibre types found in clothing. Part 2: The content of the data bank and its statistical evaluation," *Forensic Science International*, 77, 75–91.

Biermann, T.W. and Grieve, M.C. (1998) "A computerized data base of mail order garments: a contribution toward estimating the frequency of fibre types found in clothing. Part 3: The content of the data bank: Is it representative?," *Science and Justice*, 95, 117–131.

Bisbing, R. (2002) "Forensic hair comparisons," in *Forensic Science Handbook*. Englewood Cliffs, NJ: Prentice-Hall.

Connelly, R.L. (1997) "Colorant formation for the textile industry," in *Color Technology in the Textile Industry*. Research Triangle Park, North Carolina, American Association of Textile Chemists and Colorists, pp. 91–96.

Cook, R. and Wilson, C. (1986) "The significance of finding extraneous fibres in contact cases," *Forensic Science International*, 32, 267–273.

Devaney, J.M., Marino, M.A., Smith, J.K. and Girara, J.E. (2000) *Separation and Purification of Short Tandem Repeat (STR) DNA Fragments Using Denaturing HPLC (DHPLC)*. Application Note 102, Omaha, NE: Transgenomic.

Evett, I.W. (1987) "On meaningful questions: a two-trace transfer problem," *Journal of the Forensic Science Society*, 27, 375–381.

Frei-Sulzer, M. (1965) "Coloured fibres in criminal investigations with special reference to natural fibers," in *Methods of Forensic Science*, Vol. 4, ed. A.S. Curry. New York, NY: Interscience, pp. 141–175.

Gaudette, B.D. (1999) "Evidential value of hair examination," in *Forensic Examination of Hair*, ed. J. Robertson. Philadelphia: Taylor and Francis.

Gaudette, B.D. (1985) "Strong negative conclusions: a rare event," *Canadian Society of Forensic Science Journal*, 18, 32–37.

Gaudette, B.D. (1993) "Forensic fiber analysis," in *Forensic Science Handbook*, Vol. 3. Englewood Cliffs, NJ: Prentice-Hall, Inc.

Gaudette, B.D. and Keeping, E.D. (1974) "An attempt at determining probabilities in human scalp hair comparison," *Journal of Forensic Sciences*, 19, 599–606.

Holinski-Feder, E., Muller-Koch, Y., Friedl, W., Moeslein, G., Keller, G., Plaschke, J., Ballhausen, W., Gross, M., Baldwin-Jedele, K., Jungck, M., Mangold, E., Vogelsang, H., Schackert, H.-K., Lohse, P., Murken, J. and Meitinger, T. (2001) "DHPLC mutation analysis of the hereditary nonpolyposis colon cancer (HNPCC) genes hMLH1 and hMSH2," *Journal of Biochemical and Biophysical Methods*, 47, 21–32.

Houck, M.M. (1999) "Statistics and trace evidence: the tyranny of numbers," *Forensic Science Communications*, 1(3), www.fbi.gov.

Houck, M.M. (2003) "Inter-comparison of unrelated fiber evidence," *Forensic Science International*, 135(2), 146–149.

Houck, M.M. and Budowle, B. (2001) "Correlation of Microscopic and Mitochondrial DNA Hair Comparisons," *Journal of Forensic Sciences*, 47(5): 964–967.

Houck, M.M. and Budowle, B. (2002) "Correlation of Microscopic and Mitochondrial DNA Hair Comparisons," *Journal of Forensic Sciences*, 47(5): 964–967.

Houck, M.M. and Koff, C.M. (2000) "Racial assessment in hair examinations," *Proceedings of the 9th Biennial Scientific Meeting of the International Association for Craniofacial Identification*, Washington, DC, July 24–28.

Liu, W., Smith, D.I., Rechtzigel, K.J., Thibodeau, S.N. and James, D.C. (1998) "Denaturing high performance liquid chromatography (DHPLC) used in the detection of germline and somatic mutations," *Nucleic Acids Research*, 26(6), 1396–1400.

Miller, K.A. (2002) "Identifying those remembered," *The Scientist*, June 10, 2002, 41–42.

Oldenburg, J., Ivaskevicius, V., Rost, S., Fregin, A., White, K., Holinski-Feder, E., Muller, C.R. and Weber, B.H.F. (2001) "Evaluation of DHPLC in the analysis of hemophilia A," *Journal of Biochemical and Biophysical Methods*, 47, 39–51.

Roberts, P.S., Jozwiak, S., Kwiatkowski, D.J. and Dabora, S.L. (2001) "Denaturing high-performance liquid chromatography (DHPLC) is a highly sensitive, semi-automated method for identifying mutations in the TSC1 gene," *Journal of Biochemical and Biophysical Methods*, 47, 33–37.

Roux, C. and Margot, P. (1997) "An attempt to assess the relevance of textile fibres recovered from car seats," *Science & Justice*, 37, 25–30.

Rudram, D.A. (1996). "Interpretation of scientific evidence," *Science and Justice*, 36(3), 133–138.

Siegel, J. and Houck, M.M. (2001) "Forensic textile fiber analysis," in *Forensic Sciences*, Vol. 3, ed. C. Wecht. New York: Mathew-Bender/Lexis.

SWGMAT (1999) "Scientific working group on materials analysis: forensic fiber analysis," *Forensic Science Communications*, 1(1), www.fbi.gov

Underhill, P.A., Jin, L., Lin, A.A., Mehdi, S.Q., Jenkins, T., Vollrath, D., Davis, R.W., Cavalli-Sforza, L.L. and Oefner, P.J. (1997) "Detection of numerous Y chromosome biallelic polymorphisms by denaturing high-performance liquid chromatography," *Genome Research*, 7, 996–1005.

Wilson, M.R., Polanskey, D., Butler, DiZinno, J.A., Replogle, J., Budowle, B. (1995) "Extraction, PCR amplication and sequencing of mitochondrial DNA from human hair shafts," *Biotechniques*, 18, 662–668.

CHAPTER 6

CEREAL MURDER IN SPOKANE

William M. Schneck
Microanalysis Section, Washington State
Patrol Crime Laboratory, Spokane, Washington

INTRODUCTION

In February of 1999, the residence of James Cochran (all names have been changed) was found engulfed in flames. After the blaze was extinguished, it was determined that Kevin, the 11-year-old son of James Cochran was missing. Cochran made a less then convincing plea on the local news, asking for his son Kevin to return, even if he had been playing with matches and started the fire. Kevin's backpack and in-line skates were found east of Spokane along a road approximately 15 miles from his home. Two days after the fire, the fully-clothed body of Kevin Cochran was found down a snowy embankment along a lake road north of Spokane, by the driver of a snowplow as he stopped for a short break. That same week, Cochran was arrested for embezzling funds from his employer. The forensic evidence in this case involved the analysis of numerous types of trace evidence, which included gastric secretions (vomit), soil, metal turnings, building materials, paint, and soot. The association of the large variety of circumstantial trace evidence played a vital role in the investigation of this homicide. As the victim and the suspect lived in the same home and there was no physical evidence of bloodshed at the time of death, DNA would not be a contributing factor in this case.

THE COLLECTION OF PHYSICAL EVIDENCE

After the fire at the Cochran residence was extinguished, an arson investigator from the Bureau of Alcohol, Tobacco and Firearms (BATF) along with Sheriff's deputies and members from the Washington State Patrol Crime Scene Response Team searched the residence for physical evidence. A file cabinet containing business documents was removed from one of the bedrooms. The investigation revealed the fire had started in the vicinity of a couch in the downstairs den next to Kevin's bedroom. The cause of the blaze was reported as arson. Interestingly,

Figure 6.1

The victim was found along a lake road by a snow plow driver. The victim's shoes are tied on the wrong feet.

a neighbor reported that the Cochran's had two Rubbermaid garbage cans, but only one could be located at the residence.

Two days after the house fire, the frozen body of Kevin Cochran was found along a scenic lake road in Sierra County, fully clothed and face up, down a snowy embankment (see Figure 6.1). The shirt, pants, and jacket Kevin was wearing exhibited a large amount of creamy brown vomit. Vomit could also be seen on his face and in his mouth. Kevin's shoes were tied, but were surprisingly on the wrong feet. An oval shaped abrasion was noted around the left eye. At autopsy the medical examiner determined the cause of death to be strangulation. The boy's stomach contents, fingernail clipping, hand swabs, and clothing were collected as evidence for laboratory examination.

During an interview with a neighbor, investigators learned that Cochran spent an evening after the fire at their home and gave them a file folder containing documents, specifically the homeowners and life insurance policies of his children. In an interesting twist, while the detectives were conducting the interview, the neighbor received a phone call from Cochran while in jail. The neighbor said he was going to give the documents to the police, and Cochran's reply was, "my goose is cooked".

James Cochran was fast becoming a suspect in this case. He was arrested shortly after the fire on other charges pertaining to embezzling funds from his employer, a plumbing contractor. It was determined by transponder data, that Cochran made several cellular telephone calls the day of the fire north of Spokane in an area near the body dump location. Investigators believed that Kevin Cochran was strangled in his home by his father and transported in his pickup truck to the lake road dumpsite. The motive investigators believed, was money from the life insurance policy safely in the hands of a neighbor.

Figure 6.2
Robert Cochran's pickup truck. Arrows point to the front tow hook and license plate bracket showing soil and vegetation.

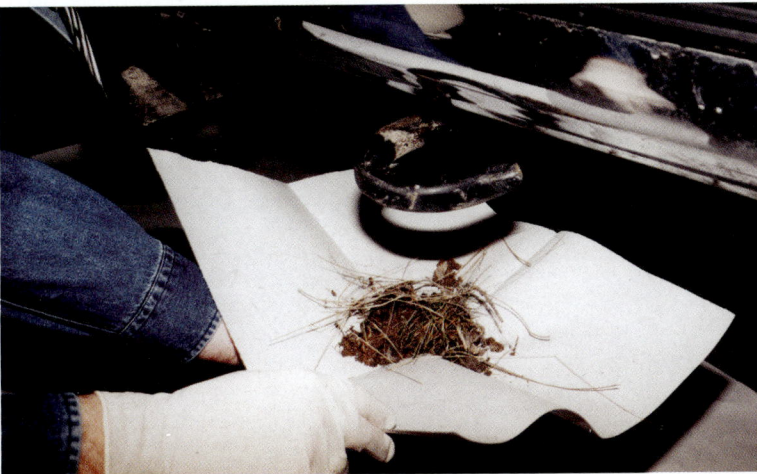

Figure 6.3
Soil and vegetation recovered from Cochran's vehicle.

A full-size Chevrolet pickup truck that James Cochran was driving on the day of the homicide was seized and searched for physical evidence. The front fender and a tow hook on the pickup truck contained several areas of crushed rock, soil, fresh grass, and pine needles. The front license plate and plate bracket had sustained what appeared to be recent damage, with imbedded soil and vegetation. These soil samples were carefully collected to preserve any layering that may have been present in the soil (see Figures 6.2 and 6.3). The black plastic bed liner was searched for trace evidence. Several droplets of light brown to pink material were observed on the driver's side wheel well hump, and in various locations on the mid-portion of the bed liner (see Figures 6.4–6.6). The scientist collecting these droplets noted the smell of possible vomit while scrapping to recover the stains.

The exterior of a Mazda Miata that had been parked in the Cochran's two-car garage at the time of the fire was searched. During the suppression of the

168 TRACE EVIDENCE ANALYSIS

Figure 6.4
The bedliner of the suspect's pickup truck. Arrows point to suspected vomit stains.

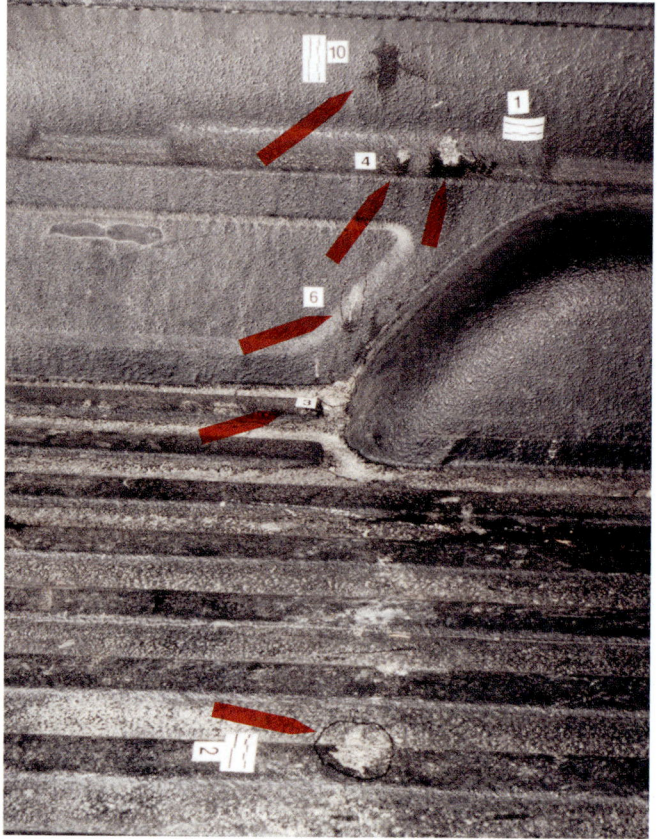

Figure 6.5
A close-up view of Cochran's pickup truck bedliner with arrows pointing to suspected vomit stains.

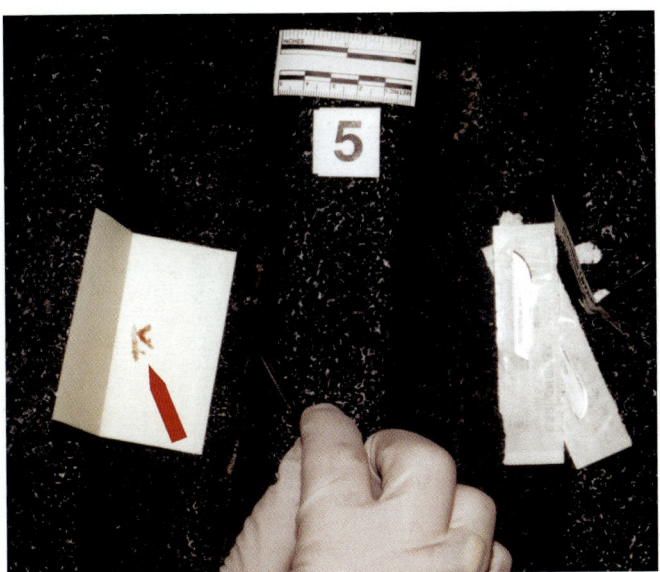

Figure 6.6
The collection of particles from the bedliner of the suspect's truck. Arrow points to a Post-It Note© containing suspected vomit.

blaze, firefighter's pushed the vehicle into the driveway. Neighbors then pushed the vehicle into a garage across the street from the Cochran residence. Upon examination, a droplet that had the appearance of vomit was located on the vehicle's passenger side rear fender and was collected. Soot was observed covering the stain, indicating the stain was applied to the vehicle prior to the fire (see Figure 6.7). During the search for physical evidence, it was of paramount importance to accurately reconstruct the chain of events. The microscopic trace evidence would prove vital in this regard.

LABORATORY EXAMINATION OF THE TRACE EVIDENCE

VOMIT AND CEREAL

Investigators were interested in the stains observed in James Cochran's pickup truck. They remembered that some of the stains may have smelled like vomit during the collection process, and wanted to know if these stains could be vomit originating from Kevin at or during the time of his death. It was essential to be able to identify the particles in the stains.

A discussion of vomit would not be complete without a brief discourse on the work of William Beaumont M.D., a surgeon stationed at Fort Mackinac in 1822. The first experiments on gastric digestion were performed upon an unfortunate individual by the name of Alexis St. Martin, a Canadian voyageur. While standing in the American Fur Company at Fort Mackinac, an accidental discharge

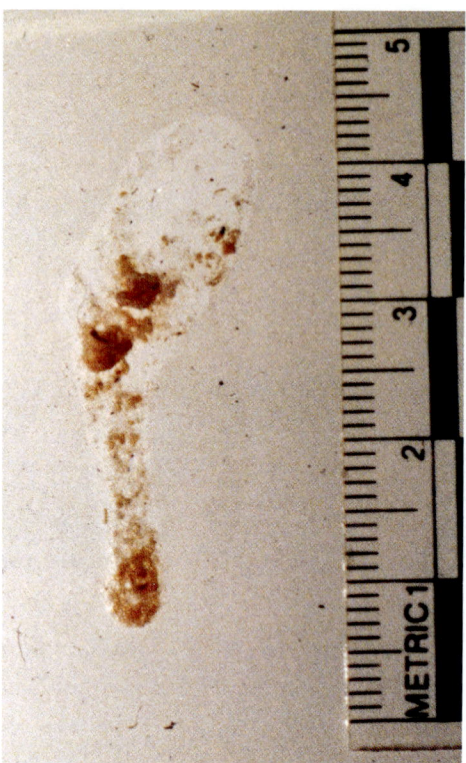

Figure 6.7
Vomit stain on the passenger side of a Mazda Miata. Black soot particles covered the stain.

from a shotgun lacerated St. Martin's diaphragm and perforated his stomach, reportedly with stomach contents oozing into the wound site. Dr. Beaumont attended to his wounds and after 3 years of convalescence began conducting digestion experiments on St. Martin's still open stomach wound. Beaumont inserted different foods directly into St. Martin's stomach in cloth sacks with an attached string, and recorded the time taken for the food to digest in his stomach. In 1833, Beaumont published Experiments and Observations on the Gastric Juice and the Physiology of Digestion, a book based upon his experiments on St. Martin. His observations laid the groundwork for future study on the human stomach and the digestive process (Gantner, 1976).

The process of digestion begins in the mouth where enzymes in saliva begin to break down starch products as food is chewed. Food is then swallowed and travels to the stomach, where the physical action of peristalsis churns and kneads the food into a semi-solid amorphous mass called chyme. Gastric juices secreted from the walls of the stomach add hydrochloric acid, mucus and gastric enzymes pepsin and renin. These enzymes start the breakdown of proteins into amino acids. Renin plays a role in the stomach of the infant. It curdles milk and allows the pepsin to work. Unless provoked, food normally resides in the stomach for up to 6 hours. When an unfortunate individual becomes ill or is violently attacked, physiologic reactions may occur which discharge food from the stomach and out through the mouth.

Although gastric fluid stains (vomit) are uncommon at crime scenes, they need to be evaluated for forensic significance. The examination of vomit may assist in criminal and civil investigations in several ways. As revolting as it may first appear, the examination of food particles found in vomit may reveal culinary secrets from an individuals last meal. Microscopically searching through traces of vomit may reveal particles of tomato, lettuce, sesame seeds, and muscle, thus suggesting a "fast food" hamburger meal. The presence of food ingredients, such as curry, bamboo shoots, etc., may help an investigator pinpoint a particular restaurant the victim may have last dined. An unusual food ingredient or an abundance of one type of food may suggest traits or food habits helpful to an investigation. The examination of vomit stains may prove or disprove a suspect or victim alibi, support witness testimony and help in crime scene reconstructions. The transfer of vomit from one individual to another cannot be overlooked. Vomit may initially transfer from a victim to a suspect environment, i.e. clothing, dwelling or automobile. When stomach contents from a victim and foreign samples from the suspect environment are characterized and compared, similarities in food ingredients may suggest a common origin linking the suspect to the victim. The mere presence of vomit when none is reported in an alibi statement may be important to an investigator.

The identification of vomit stains is based on the characterization of partially digested food ingredients and the presence of gastric enzymes. Samples submitted to the forensic laboratory may include stomach contents collected at autopsy, dried stains collected at crime scenes, or stains found on materials such as clothing. The presence of drips, projected patterns and transfer stains are important to document. Vomit stains are often recognized by the all too familiar gross appearance and attendant unpleasant odor. But further scientific inquiry is often required to confirm a given stain is indeed vomit. During the crime scene investigation, vomit stains are documented in notes, sketches, and photographs. A dried stain can be collected by scraping and particle picking with a forceps to a small paper envelope. In the laboratory the suspected vomit stain is examined visually and documented as part of the routine clothing or materials examination. An ultra-violet light can cause some vomit stains to glow, allowing rapid detection, but this type of illumination may have deleterious effects upon nucleated cells useful in DNA analysis. When a stain has been located visually, a stereobinocular microscope can be used to magnify the stain up to approximately 100 times. Individual particles within the stain can be particle picked to a microscope slide for further analysis by polarized light microscopy (PLM). A portion of the stain can be removed and tested for the presence of gastric enzymes common in gastric fluids.

TESTING FOR GASTRIC ENZYMES IN VOMIT STAINS

The literature contains only a few references concerning the identification of gastric fluid, and in particular vomit stains. Simon (1897) describes a test for the presence of gastric enzymes. If a specific quantity of gastric juice coagulates in the presence of milk, the enzymes chymosin (also known as renin) and chymosinogen are present. The milk coagulating activity of human gastric fluid has been attributed to pepsin. The only published study dealing with gastric enzymes in a forensic context is by Lee and Gaensslen (1985). They modified clinical assay procedures to test for the pepsin and renin-like activity on gastric fluid and gastric fluid stains. The author repeated some of Lee and Gaensslen's work by testing saliva, semen, urine, feces, whole blood, and many foods for the presence of coagulation in whole milk. None of the materials tested coagulated whole milk. Pepsin is a proteolytic enzyme found in the gastric secretions of many vertebrates. Proteolytic enzymes hydrolyze or break down proteins or peptides into simpler more soluble products during the digestion process. The presence of gastric enzymes in a stain indicates it originated from a mammal, but not necessarily a human.

A quick screening test for the presence of gastric fluid can be used to identify stains which are suspected to be vomit. Known vomit samples as small as 0.5 mm in diameter have tested positive using this procedure. A small portion of the suspected stain is removed to a black porcelain spot plate along with a dried vomit control. Several drops of whole cows milk are pipetted into each well. The spot plate is placed into a humidity chamber at 38°C for approximately 30 min. The spot plate wells are then examined visually and with a stereobinocular microscope. The occurrence of coagulation or curdling indicates gastric enzymes are present in the sample (see Figure 6.8). The known vomit sample should also exhibit coagulation. The reactions on the spot plate are then documented photographically. The presence of gastric enzymes was not tested in the Kevin Cochran evidence. The test was perfected after the case was completed.

IDENTIFICATION OF FOOD TRACES BY LIGHT MICROSCOPY

With limited training many examiners could compare two unknown particles and describe similarities and differences, but to access uniqueness, commonality, and unusual features, one must be able to accurately identify the particles in question, the signature of a true microscopist.

Everyone can recognize and differentiate a dinner roll from a French fry when they are whole or even sliced. But how could you tell them apart if they have been masticated, and partially digested in the stomach, then regurgitated onto fabric and allowed to dry?

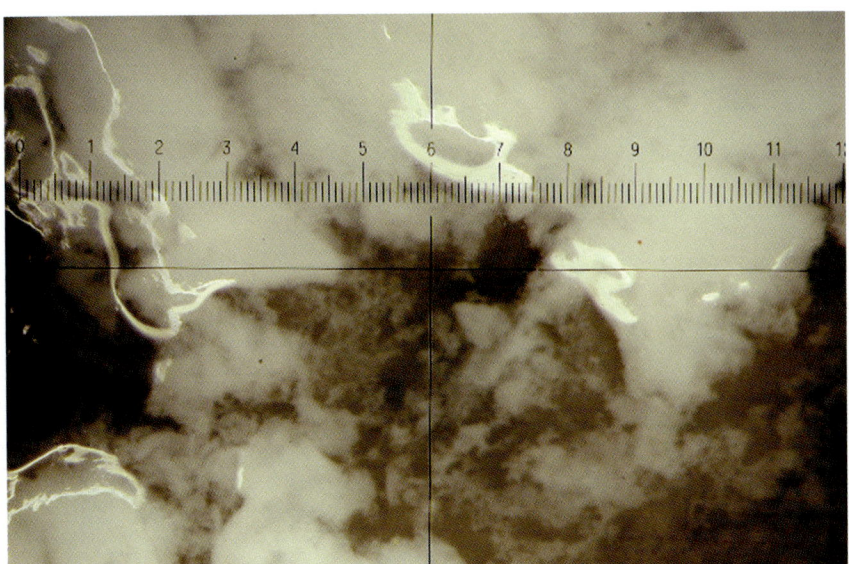

Figure 6.8
A known vomit sample curdled in the presence of whole milk, a positive reaction for the presence of gastric enzymes.

The stereobinocular microscope and the polorized light microscope (PLM) are the instruments of choice in the characterization and identification of food particles. No other instruments are as versatile in identifying materials. The theory, techniques, and applications in PLM are resplendent in the scientific literature, and can be found elsewhere. The informative textbooks, *Food Microscopy* (Flint, 1994), *Identifying Plant Food Cells in Gastric Contents for use in Forensic Investigations: A Laboratory Manual* (Bock, 1988), *The Particle Atlas* (McCrone and Delly, 1973), *The Particle Atlas* (McCrone, Delly and Palenik, 1979) and the *Atlas of Microscopy of Medicinal Plants Culinary Herbs and Spices* (Jackson and Snowdon, 1990) are excellent references available when characterizing and identifying food particles.

The stereobinocular microscope can be used to examine vomit stains in situ, showing the relationship of food particles as they were applied to the substrate, i.e. clothing (see Figure 6.9). A PLM has many of the same features as a common biological microscope, such as a binocular head with two eyepieces, several objectives on a rotating nosepiece with a magnification range from 4× to 100×. Additionally the PLM is equipped with a round rotating stage, and two polarizing filters, one below the stage (the polarizer) and one above the stage (the analyzer). An accessory slot in the microscope allows the insertion of waveplates and other more specialized apparatus for optical characterization. The PLM may also be fitted with phase contrast objectives and a specialized phase condenser lens. Phase contrast is a useful method to improve contrast when examining materials, such as food plant cells that have similar refractive indices between grain boundaries. The PLM can magnify and resolve small particles down to the micrometer level, and determine the many optical characteristics used in identification.

Figure 6.9
A vomit stain from Kevin Cochran's shirt.

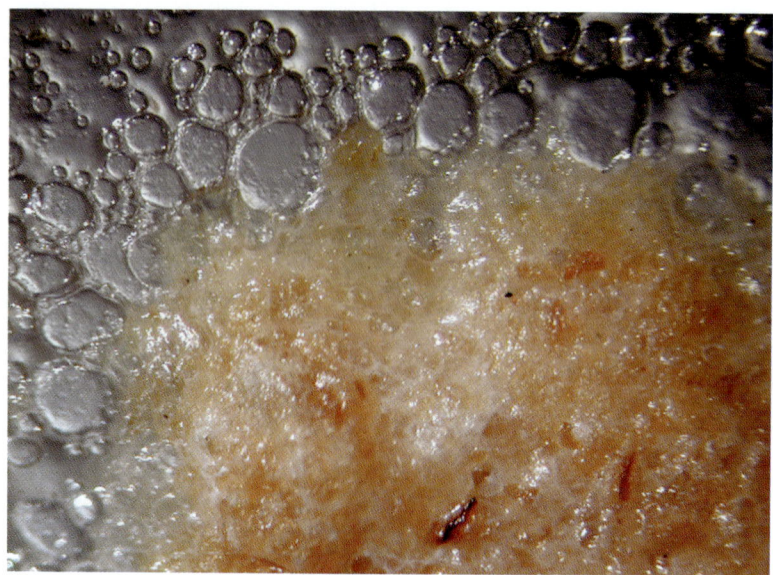

There are a variety of ways to prepare gastric fluids for microscopical examination. If the stomach contents are received in the liquid state they can be examined directly by applying a smear to a microscope slide with the addition of a coverglass. Individual particles can be removed from the fluid by particle picking to a microscope slide with a needle or forceps and applying a small drop of distilled water to disperse the particles. A subsample of the gastric contents can be wet sieved with distilled water through a variety of fine mesh screens which will both clean and size separate the particles. A drop of the fluid portion of the vomit can be gently smeared across a microscope slide and a cover glass applied over the smear for microscopic examination. Bock (1988) describes a procedure to prepare both temporary and permanent slides of gastric contents. If a dry vomit stain is received, a portion of the stain can be scrapped to microscope slide and resuspended in distilled water or a variety of other mounting media, such as Norland Optical Adhesive, Cytoseal, Cargille liquids, or Permount.

Stains can be used to improve contrast and aid in the identification of food products. Some of the more common stains used in the examination of food products include Toluidine Blue, aqueous iodine solutions, Safranin, Trypan Blue, and Oil Red "O" (see Figures 6.10 and 6.11) (Shane, McCrone Research Institute).

The microscopist's observation of thousands of particles through years of casework and research will develop a visual "cerebral encyclopedia". The morphology, color, and internal structure of a given food can be translated rather quickly into known food groups. The microscopist should have available dichotomous plant cell keys (Bock, 1988), commercially prepared microscope

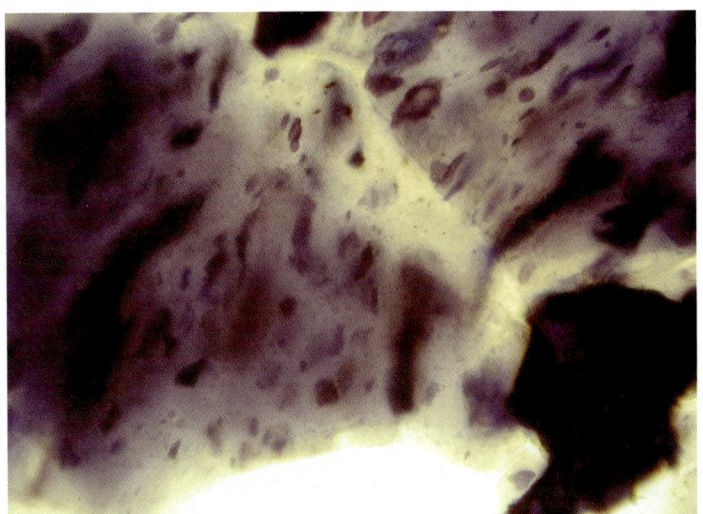

Figure 6.10
Processed starch stained with Trypan Blue.

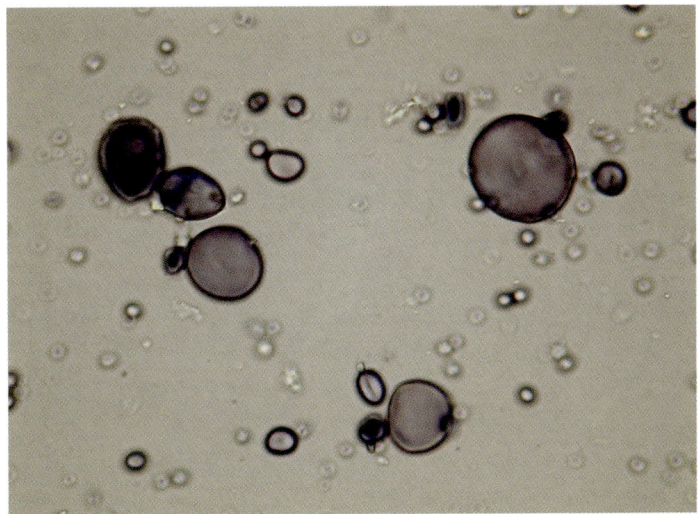

Figure 6.11
Unprocessed wheat starch stained using an aqueous solution of iodine and potassium iodine (plane polarized light).

slide sets and individually prepared food standards including various starches, prior to undertaking the identification of food particles.

Vomit may contain food particles from meat, grains, vegetables, dairy products, fruits, nuts, and fats. Large fragments of food, such as seeds, pieces of meat, and leafy vegetables can often be identified visually and with a stereobinocular microscope. Smaller food particles that have been partially digested require closer scrutiny using the PLM.

Starch, a carbohydrate storage product found in all chlorophyll containing plants, is common in many foods. Chemically, starch is composed of two types of glucose polymer: amylose and amylopectin. Starch occurs as insoluble granules stored primarily in the roots, tubers, and seeds of plants, but may also occur in

Figure 6.12
Potato starch (crossed polarized light).

Figure 6.13
Corn starch (partial crossed polarized light).

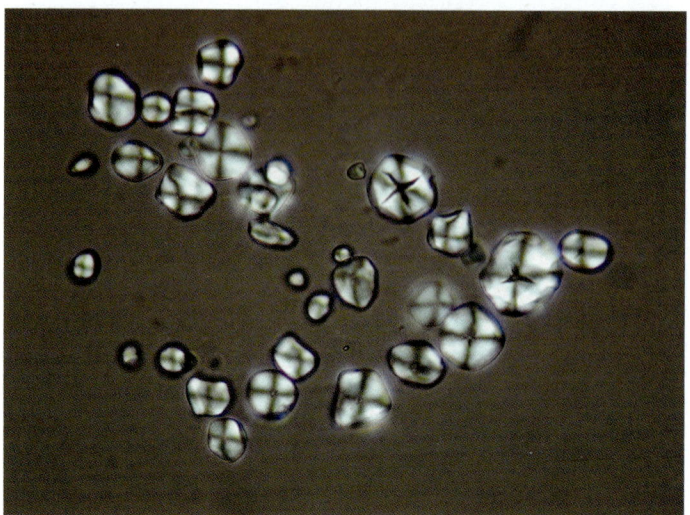

the leaves and stems. Important commercial sources of starch are derived from the cereals (corn, wheat, and rice) and the root tubers (potato and cassava-tapioca). The starch granule develops stratified layers which form around a nucleus, called the hilum. The size of starch grains can be readily measured and compared to known standards of starch using the PLM. The shape of starch grains can vary from near perfect spheres to polygons, flattened spheroids, elongated disks and many others. In crossed polarized light, unprocessed starch grains have a characteristic "Maltese cross" extinction pattern, with the hilum often near the center of the cross. The size and shape of each starch grain is characteristic of the plant in which it is derived (see Figures 6.12–6.15).

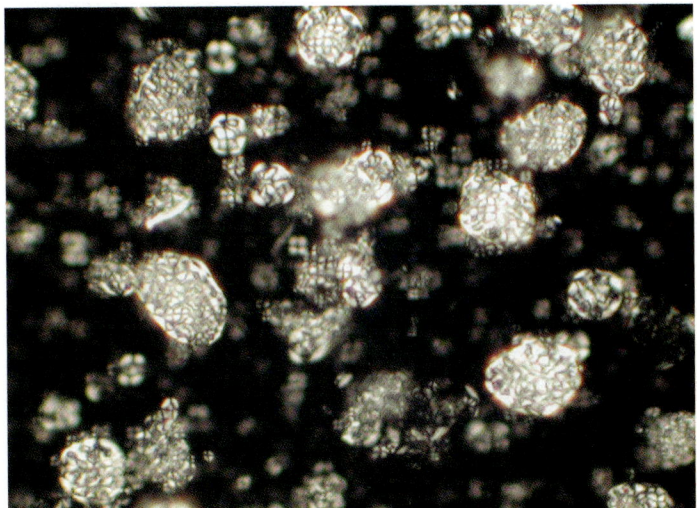

Figure 6.14
Oat starch (crossed polarized light).

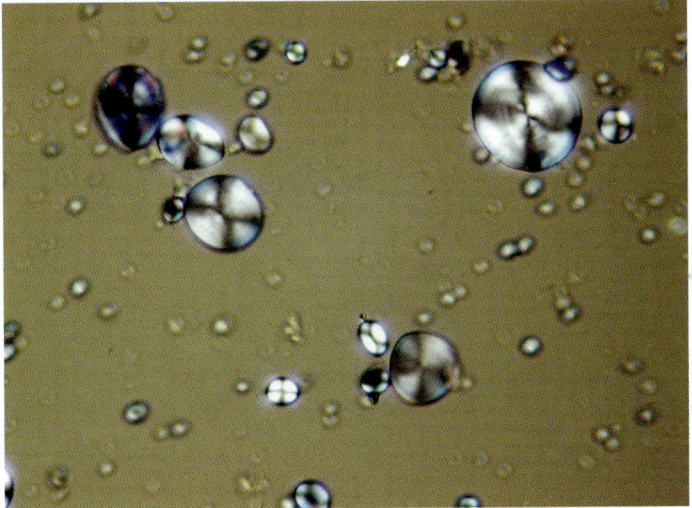

Figure 6.15
Wheat starch from Pillsbury® All Purpose flour (iodine-stained in partial crossed polarized light).

Starch can be found in food in the raw, chemically modified or pre-gelatinized form. Aqueous iodine staining tests for starch have been developed, most of which are composed of a solution of iodine, potassium iodine, and distilled water. The different amounts of amylose and amylopectin in a starch affect its reaction to testing with iodine solutions. High amylose starches stain blue with an aqueous iodine solution, whereas starch high in amylopectin will stain yellow to a dull reddish color.

The microscopic examination of cooked or processed starch presents a greater challenge to the microscopist. At elevated temperatures in the presence of water, starch undergoes a process called gelatinization. This process can be

Figure 6.16
The vomit-stained shirt worn by Kevin Cochran.

observed microscopically beginning with granule swelling and progressing at higher temperature to the loss of the Maltese cross birefringence pattern observed in polarized light. The gelatinization temperature range for known starch types is well documented whereby an unknown grain can be compared with some success. The stain Trypan Blue is recommended for the examination of processed starch. Trypan Blue will only stain damaged or cooked grains, leaving the intact uncooked grains clear.

Stains from the bed of the pickup truck and from the exterior of the Mazda Miata were compared to the vomit and gastric contents of Kevin Cochran for similarities. Initial examination of Kevin's shirt revealed light brown to pink "vomit" smears extending across the front of the shirt (Figure 6.16). During questioning one of Kevin's sisters, it was discovered that he was last seen eating cereal in the kitchen the morning of the fire. Investigators returned to the residence and recovered containers of cereal from the kitchen. Two plastic containers, one of which appeared to have a "cocoa puffs"-type cereal, and another with cereal particles consisting of light brown anchor-shapes, and irregular shaped red, yellow, pink, orange, and blue grains were collected. Two opened and partially consumed plastic bags labeled; Apple Cinnamon Toastyo's®, and Marshmallow Mateys® were also submitted. The anchor-shaped cereal was visually similar to the Marshmallow Mateys® brand in the plastic bag (see Figure 6.17).

If the cereal found in the kitchen of the Cochran residence "matched" the cereal in the vomit on Kevin's clothing, and was found to be similar to stains in the pickup truck, investigators may have a connection linking James Cochran to the death of his own son.

Cereals are actually members of the grass family *Gramineae*, and include such notable members as wheat, rye, barley, maize, rice, oats, and millet. The name

Figure 6.17
Marshmallow Mateys® breakfast cereal. The anchor-shaped particles contain oat flour whereas the colored particles contain processed corn starch and sugar.

cereal is derived from the word Ceres, the Italian goddess of agriculture. Particle identification of specific plant parts requires knowledge of plant botany. Some of the best resources are early 20th century vegetable food and pharmacognosy textbooks (Greenish, 1923; Winton, Moeller and Winton, 1916; Winton and Winton, 1932). Illustrations from these early texts show detailed microscopic structures, many of which have more information than current photomicrographs.

Examination of the cereal brands from the residence revealed three distinct types; Marshmallow Mateys®, Apple Cinnamon Toastyos®, and a "Cocoa Puffs"-like cereal. Numerous microscope slides from the three different cereals were prepared for examination and comparison by PLM. Individual particles of different shapes, colors, textures, and sizes were crushed and mounted in water, Cargille high dispersion liquids, and Norland Optical Adhesive.

All three cereal brands could be distinguished microscopically. Furthermore, microscopic examination of particles in all submitted vomit stain samples revealed the presence of cereal grain products with microscopic similarities to that observed in the Marshmallow Mateys®, and dissimilar to those observed in Apple Cinnamon Toastyo's® and the "Cocoa Puffs" cereal.

The ingredients label on the bag of Marshmallow Mateys® stated the cereal contained; whole oat flour (including the oat bran), marshmallow bits (sugar, corn syrup, modified corn starch, dextrose, gelatin, artificial flavor, and colors), sugar, corn syrup, wheat starch, salt, trisodium phosphate, calcium carbonate,

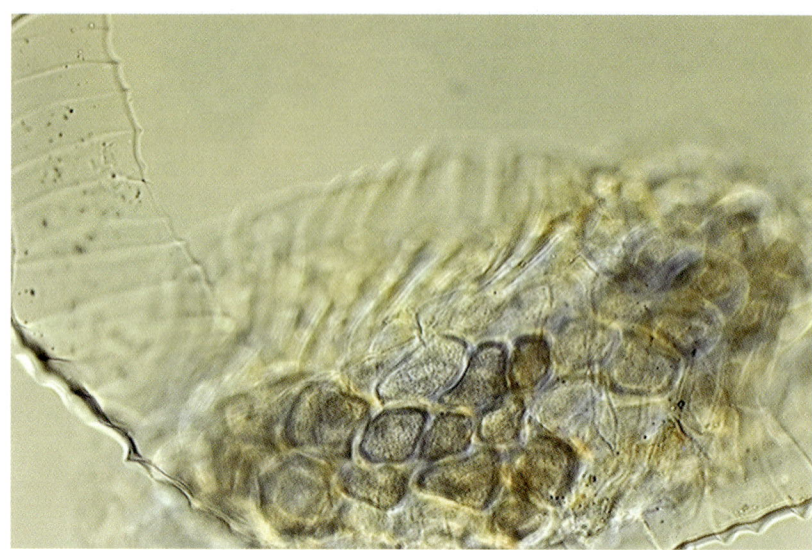

Figure 6.18
Microscopic structures in oat flour. The lower portion of the photomicrograph shows aleurone cells in the bran coat. The delicate, colorless, spermoderm cells can be seen above and below the aleurone cells.

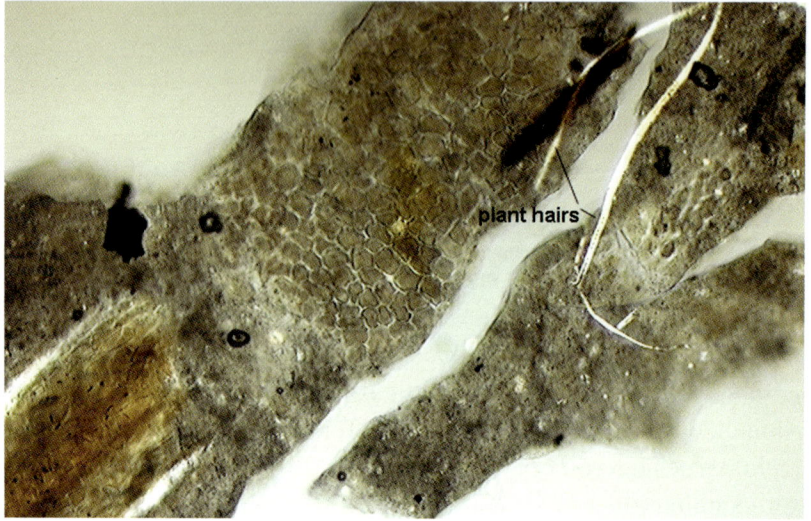

Figure 6.19
Microscopic structures in vomit containing oat flour. The elongated fibrous structures are plant hairs, also called trichomes, common in oat flour.

sodium ascorbate, and additional vitamins and minerals. The term flour denotes a grain that has been ground to a desired consistency. Oat flour is used in extruded cereal products, cookies, and crackers. PLM in conjunction with a few micro-chemical and color tests identified a majority of the ingredients. Common microscopic structures observed in oats included the outer epidermis composed of elongated cells exhibiting hairs, the parenchyma containing distinctive starch grains, the paper-thin inner epidermis with hairs, elongated cells, and stomata (see Figures 6.18 and 6.19). Sugar, confirmed using the Anthrone micro-color test, was observed on the surface of the marshmallow bits (Figure 6.20) and in the vomit samples.

Figure 6.20
*A Marshmallow Mateys®
marshmallow bit.*

In this particular case, starch observed in Kevin's stomach contents, questioned vomit stains and known cereal products was tested using Lugol's Iodine solution (1 g iodine, 2 g potassium iodine, and 300 cc distilled water) and Trypan Blue (0.25 g Trypan Blue mixed with 100 cc distilled water). Iodine stains starch granules blue to purple-blue dependent upon the strength and time allowed for the stain to react with the starch granules. Trypan Blue was used to stain for damaged and gelatinized starch. Trypan Blue will leave intact, unprocessed starch unstained so that grain birefringence characteristics can be observed. Oil Red "O" was also used to stain for liquid fats present in the food and vomit. Unprocessed oat and wheat starch and gelatinized starch were identified in the vomit samples.

A microscopical examination and comparison of stains found on the pickup truck bed liner and a stain on the exterior of the Mazda Miata revealed the presence of vomit with cereal ingredients similar to that found in the vomit on Kevin's clothing and gastric fluid. The cereal ingredients were consistent with Marshmallow Mateys®, the final meal of Kevin Cochran.

ADDITIONAL TRACE EVIDENCE EXAMINATIONS

SOIL

During an interview with James Cochran, investigators asked if the damage to the front fender of his vehicle was recent. Cochran stated that indeed it was, "I ran my truck off the road into some dirt on Chatsburg Road". To test the validity of his story, soil, rock, and vegetation controls were collected along the 3 mile stretch of Chatsburg Road in Spokane at ¼ mile intervals. Soil control

Figure 6.21

Fine sand-size minerals wet sieved from soil recovered from Cochran's pickup truck (partial crossed polarized light).

samples were also collected along the lake road where Kevin's body was discovered and along the roadway where his back pack was found.

Soil samples from the vehicle contained clumps of fresh vegetation with attached topsoil, Ponderosa pine needles, and twigs with attached lichen. The front license plate and its bracket were damaged and distorted with lodged pieces of crushed granitic rock fragments and soil. The damage to the front of the vehicle indicated it may have made contact with a grassy rock embankment or a rock outcrop composed of granite. The 3 mile stretch of Chatsburg Road did not have exposed rock outcrops. There was no physical evidence of a vehicle swerving off the road into a ditch or embankment along that stretch of road. Soil samples from the vehicle were sieved with distilled water and the fine sand-size grains examined and identified by PLM (see Figures 6.21 and 6.22).

The fine silt and clay portion of the soil was dried and its soil color compared to the soil along Chatsburg Road using the Munsell Soil Color Charts. The soil from the pickup truck contained distinctive phytoliths, not observed in the soil from the alleged location he said he ran off the road (see Figure 6.23). Phytoliths are particles of hydrated silica derived from the cells of living plants that have been liberated from the cells upon death and decay of the plant. Phytoliths with taxonomic significance are produced in great numbers by many families of higher plants. Due to their chemical structure, phytoliths are retained in soil over long periods of time (Piperno, 1988). Phytoliths are only one example of a multitude of microscopic particles, both organic and inorganic, that can be identified and compared in soil. The soil on the pickup truck also differed from the soil along the lake road where Kevin's body was recovered. Soil along the highway where the backpack was found consisted of particles from a granitic source with petrographic similarities to soil particles

Figure 6.22

Fine sand-size minerals wet sieved from soil recovered from Cochran's pickup truck (partial crossed polarized light). The green mineral is a pleochroic amphibole.

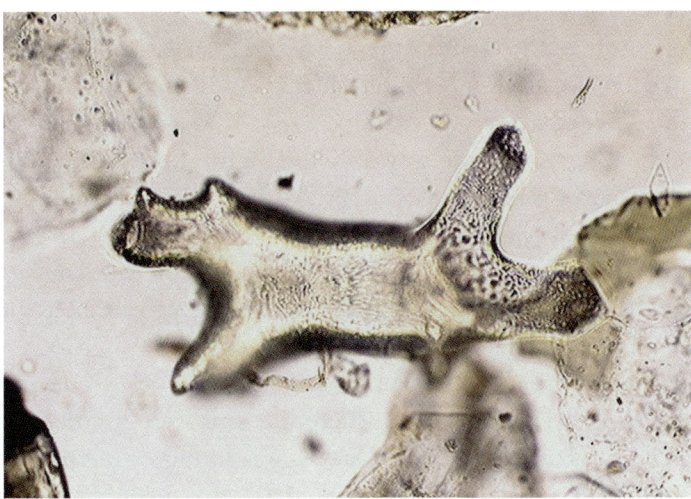

Figure 6.23

A phytolith in a soil sample from Cochran's pickup truck (plane polarized light).

on the pickup truck. Soil on the pickup truck was dissimilar to the soil from the location he said he ran off the road, and more consistent to soil along the highway where his sons backpack was recovered.

GARBAGE CANS

Investigators now had solid physical evidence that demonstrated how Kevin was transported in the bed of his father's pickup truck. Investigators were interested in how the boy's body was concealed and transported in an open pickup truck during the morning hours after the house fire. They remembered that one Rubbermaid garbage can was missing from the home. Could Kevin's body have

Figure 6.24 Rubbermaid® garbage cans with identical physical characteristics. The can on the left was found in the Cochran garage. The can on the right was recovered by a well-meaning citizen from a gas station lot.

been placed into a garbage can to conceal his final trip? An 11-year-old boy with physical characteristics similar to Kevin was asked to crawl into a garbage can like the one at the Cochran residence. Indeed, you could fit an 11-year old inside the garbage can.

Investigators made a plea on the nightly news for information leading to the recovery of a blue Rubbermaid garbage can like the one at the Cochran residence. This plea did not go unanswered. At least 15 similar appearing blue garbage cans were seized as evidence by the Spokane Sheriff's Office. The daunting task of examining the cans for trace evidence that could link Kevin's body to a garbage can was underway.

Most garbage cans could be eliminated by class characteristic dissimilarities. Locations were given as to where the cans were found. One can with matching physical characteristics and labels, was found on a rural highway not far from where Kevin's backpack was found (Figure 6.24). The examination of this can revealed white paint droplets on the lid that appeared to be overspray from a paint roller. The can found at the Cochran residence also had similar appearing paint overspray. When the lid was opened light brown to pink stains that looked similar to the Marshmallow Mateys® vomit were observed (Figure 6.25)! Further microanalysis confirmed that the stains were indeed vomit and that they were microscopically similar to the vomit from Kevin Cochran. Analysis of the white droplets on both cans was conducted using PLM and scanning electron microscopy, coupled with an energy dispersive spectrometer. These examinations were conducted to identify the inorganic ingredients in the white paint droplets. White primer paint and gypsum-containing wall texture particles were found on both garbage cans. The particles contained diatomaceous earth, titanium dioxide, and clay. These finding suggested that both cans were located in approximately the same

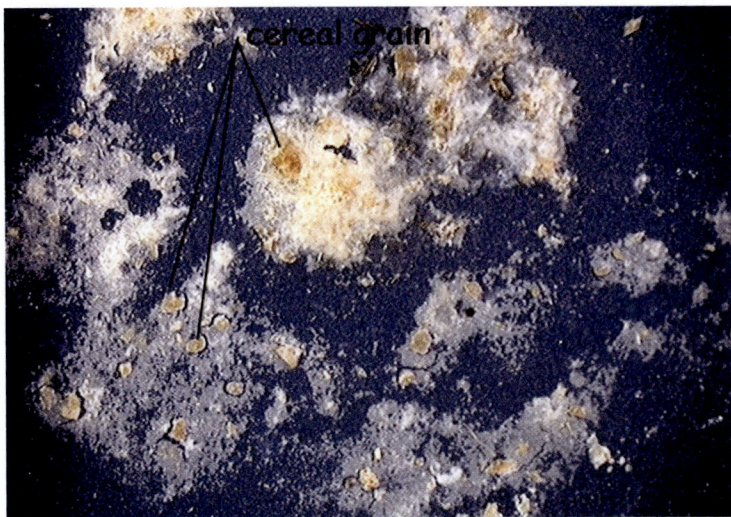

Figure 6.25
Vomit with cereal grain particles found on the inside of the Rubbermaid garbage can.

place during application of paint and wall texture. Later investigations revealed that a workshop on the Cochran property was recently finished and the walls painted. The interior of the garbage can was cut open and all materials examined. As expected, a vast assortment of debris was found in the can including paper, food wrappers, soil, a tampon, aluminum foil, and hairs. Fine particles adhering to the side walls and bottom of the can were brushed out and wet sieved with distilled water through a set of brass screens of progressively smaller size. This process was used to both clean and size separate the material prior to characterization. Materials identified in the can included metal turnings, iron spheres, rust, charcoal, blue, red and orange architectural paint chips, and glitter particles. Blue, green, gold, clear, red, and silver glitter particles, approximately 1 mm in size, exhibited hexagon, square, and irregular-shapes. Glitter particles found in Kevin's backpack were similar to those found in the garbage can (see Figure 6.26).

The oval abrasion found around Kevin's eye had similarities in shape to a molded protruding structure in the bottom of the garbage can. Could Kevin have been placed face down in the garbage can?

The pants, shirt, and jacket Kevin Cochran was wearing at the time of his death were examined for trace evidence which could place him inside the garbage can. The shirt and pants were brushed down over fresh examination paper to collect any particles adhering to the clothing. The material found on the pants, shirt, and coat included black iron spheres, metal turnings, blue, red and yellow paint chips, and a green glitter particle.

Hand swabbings collected at autopsy were examined and found to contain metal turnings similar to those found in the garbage cans, and on Kevin's clothing (see Figure 6.27).

Figure 6.27
Metal particles adhering to the inner wall of the Rubbermaid garbage can.

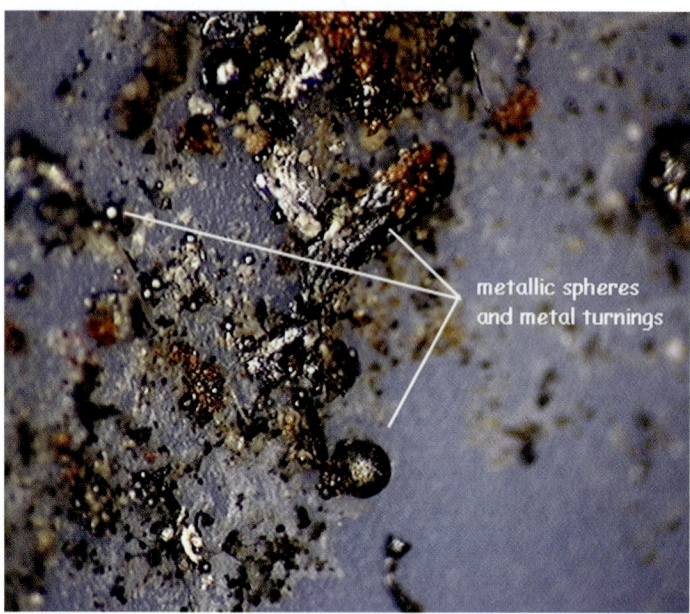

Figure 6.26
Glitter particles common to Kevin Cochran's backpack and garbage can. The particles are approximately 1 mm in size.

SOOT

Investigators were in search of physical evidence suggesting motive and or premeditation in the case against James Cochran. Cochran had an opportunity to walk through his gutted, soot filled home shortly after the fire. An examination of the documents once held at the neighbor's home was conducted to determine if they were free of soot or fire damage. The absence of soot on these documents would suggest they were purposefully removed from the house before the fire to prevent their loss and be retained to collect life insurance money from the death of his son. The absence of soot on the documents would thus bolster the claim of premeditation. An initial visual examination of the file folder and its contents revealed that they were free of soot damage. An examination of the file cabinet revealed file folders identical in construction to those given to the neighbor. Examination of the entire contents of the file cabinet was conducted to see if all the documents exhibited soot deposits, in particular the exposed file tabs on each folder.

The identification, characterization, and possible origin of soot particles from within the house was undertaken before the examination of the file folder and its documents. An examination of the exposed tabs on all of the file folders within the file cabinet revealed soot particles. After wiping off all visual soot from one of the file tabs, soot could still be detected using a Leica Stereoscan scanning electron microscope (SEM) coupled with an Oxford energy dispersive spectrometer. Therefore, even if Cochran cleaned the files, soot could still be retained and identified on the documents.

Soot is a form of particulate material ranging in size from about 10 nm to 1 mm, consisting of variable quantities of inorganic solids and carbonaceous material often combined with extractable organic compounds, such as tars and resins. Soots are formed as by-products of combustion or pyrolysis of organic compounds. Carbon soot particles have been classified by microscopy into four morphological groups: char fragments, cenospheres, aciniform, and microgel. Char fragments comprise the group with the largest particle size. They are composed of carbonized wood or cellulosic material. Cenospheres are hard shiny porous or hollow carbon spheres, which form when liquid drops undergo carbonization without substantial change. Aciniform carbon particles form through deposition in the gas phase as a straight chain or branched chain morphology, ranging in size from 20 to 30 nm. The fourth category consists of spheroidial carbon of colloidal dimensions embedded in carbon or carbonaceous material (Millette, 1998).

In a house fire an unlimited variety of materials may be partially consumed ranging from wood, wallboard, furniture, clothing, bedding, liquids, resins, coatings, etc. The SEM characterization of the soot on the file tabs from the residence revealed an interlocking network of sub-millimeter size particles. Within many of these networks were gypsum crystals composed of calcium and sulfur, originating likely from wallboard. Other characteristic particles observed were char fragments with remnants of wood structure (see Figures 6.28 and 6.29).

A comprehensive examination of the file folder, file folder tab, and the enclosed documents retained by the neighbor, failed to reveal any traces of soot. It was determined that these documents were not in the file cabinet during the fire, nor were they likely in the house during the blaze.

Figure 6.28

Soot particles from a file folder tab found in the Cochran's bedroom. The elongated particles are gypsum (SEM photomicrograph).

Figure 6.29

Soot particles from a file folder tab found in the Cochran's bedroom. The particle in the center is charred wood (SEM photomicrograph).

SUMMARY

SEQUENCE OF TRANSFER EVENTS

Sets of corresponding debris between the body of Kevin Cochran, Kevin's backpack, two garbage cans, and James Cochran's pickup truck were evaluated microscopically (see Table 6.1).

Metal turnings, paint, and wall texture debris probably originated from a newly built garage behind the Cochran residence. Cochran recently finished the interior of the garage, with the installation of drywall and a fresh coat of

	Vomit with Marshmallow Mateys®	Metals	Glitter particles	White paint and wall texture
Victim	✓	✓	✓	
Arson scene	✓ Mazda			✓ Garage
Suspect's pick-up truck	✓			
Garbage can	✓	✓	✓	✓
Victim's backpack			✓	

Table 6.1

Trace evidence associations observed between physical evidence in the Kevin Cochran homicide.

paint and plaster wall texture. Cochran enjoyed restoring "muscle cars" in the new garage. Molten metal particles from high-speed grinding tools used during vehicle restoration projects, were probably swept off the floor and into the Rubbermaid garbage cans. The garbage cans were likely near a wall where they would have come into contact with overspray during the painting process.

During the strangulation and body transport events, Kevin's vomit dripped onto the outside corner of the Mazda Miata, and into the bed of the pickup truck. As his body was placed into the garbage can, some of his vomit transferred into it. Kevin's backpack which contained glitter particles was also placed into the garbage can at the same time. Metal turnings and glitter particles were transferred from the can to the boy's body and clothing.

Detectives believed that James killed his son and set the fire to his own home to collect insurance money. To validate the financial motive for the killing, and explore the possibility of premeditation, the insurance papers were examined for soot. Soot was not observed on the file allegedly held by the neighbor, thereby demonstrating James need to protect that one file and implicating him as the arsonist and the murderer of his son.

On Memorial Day, 1999, James Cochran committed suicide in his jail cell using a co-axial cable from a television set. Although the physical evidence upon which this case was prepared will never go to court, the microscopic vestiges of truth have spoken. This case clearly demonstrates the need to continually fund and train forensic scientists in the discipline of microanalysis.

REFERENCES

Bock, J.H., Lane, M.A. and Norris, D.O. (1988) *Identifying Plant Food Cells in Gastric Contents for the Use in Forensic Investigations: A Laboratory Manual*. US Department of Justice, National Institute of Justice.

Flint, O. (1994) *Food Microscopy: A Manual of Practical Methods, Using Optical Microscopy. Microscopy Handbooks 30*. Oxford, UK: Bios Scientific Publishers.

Gantner, G.E., Dwyer, J.D. and Lynch, E. (1976) *Identification of Food Materials in Gastric Contents (Emphasizing Microscopic Morphology)*. Forensic Science Foundation.

Greenish, H.G. (1923) *The Microscopical Examination of Foods and Drugs*. Philadelphia: P. Blakiston's Son and Co.

Jackson B.P. and Snowdon, D.W. (1990) *Atlas of Microscopy of Medicinal Plants, Culinary Herbs and Spices*. Boca Raton, Florida: CRS Press.

Lee, H.C., Gaensslen, R.E., Galvin, C. and Pagliaro, E.M. (1985) "Enzyme assays for the identification of gastric fluid," *Journal of the Forensic Science*, 30, 97–102.

Millette, J.R. and Few, P.W. (1998) "Indoor carbon soot particles," *The Microscope*, 46, 201–206.

McCrone, W.C. and Delly, J.G. (1973) *The Particle Atlas*, Vol. II. Ann Arbor, Michigan: Ann Arbor Publishers.

McCrone, W.C., Delly, J.G. and Palenik, S.J. (1979) *The Particle Atlas*, Vol. V. Ann Arbor, Michigan: Ann Arbor Publishers.

Piperno, D.R. (1988) *Phytolith Analysis, An Archaeological and Geological Perspective*. San Diego: Academic Press.

Shane, J.D. *Microscopy of Botanical Traces Course Notebook*. Chicago, IL: McCrone Research Institute.

Simon, Charles E. (1897) *A Manual of Clinical Diagnosis by Means of Microscopic and Chemical Methods, for Students, Hospital Physicians, and Practitioners*. Philadelphia and New York: Lea Brothers and Co., pp. 141–146.

Winton, A.L. and Winton, K.B. (1932) *The Structure and Composition of Foods*, Vol. 1. New York: John Wiley and Sons.

Winton, A.L., Moeller, J. and Winton, K.B. (1916) *The Microscopy of Vegetable Foods*. New York: John Wiley and Sons.

CHAPTER 7

USING 1:1 TAPING TO RECONSTRUCT A SOURCE

Kornelia Nehse
Textile Expert and Head of the Fibers Group
German State Police, Berlin, Germany

INTRODUCTION

One morning in April 1994, two nurses were found dead in the kitchen of a hospital for infirmed elderly people – all of the patients were confined to their beds. The nurses, a woman and a man, were found at around 5 o'clock in the morning. Both of them were tied to a chair with kitchen towels and had fallen over onto the kitchen floor, lying in a large pool of blood. Their lifeless bodies were clothed in white hospital uniforms (Figure 7.1).

The night shift began at 10:00 p.m. and nobody discovered the absence of both victims until the janitor found them in the morning and reported it to the police. The lack of witnesses resulted from the immobility of all the patients but

Figure 7.1
Victims on the kitchen floor.

also from the lack of lasting and credible memories in nearly every case – even if they had seen something, their recollections could not have been trusted. From the forensic point of view, the case was a real challenge.

When the crime scene team appeared on the scene no obvious evidence could be detected in the near environment with the unaided eye – no suspected weapon was left at the crime scene to compare, no relevant fingerprints were found. Only a few small items from a package of the type found in survival knives, like toothpicks, were found on the scene. Although the kitchen floor was covered in blood, no footprints other than those of the victims could be seen. And even stranger was that no footwear impressions, no bloody imprints of any kind, were found anywhere in the hospital corridors.

In the absence of visible physical evidence, the investigators had to rely on "invisible" trace evidence. Since 1987, murdered victims in Berlin are taped 1:1 on the crime scene, if possible, when the circumstances indicate very close contact between the victim and the offender. Therefore, this case was predestined for this procedure, considering that the crime took place in the very limited area of the hospital kitchen ($3 \times 4\,m$) and at least one culprit must have been in a very close and intensive contact with the victims and the towels from the hospital kitchen he or she used to tie both victims to the chairs. The potential to find trace evidence in the best sense of Edmund Locard's transfer theories was very high.

There was also another very important aspect: no one entered the kitchen before the crime scene team, not even the caretaker who found them. To avoid further contamination and to get proper results, the 1:1 taping has to be carried out as the first measure at a scene. This is extremely important, as Ryland and Houck pointed out "Conclusions concerning its (trace evidences) significance are often not individualizing; that is, there is usually more than one potential source of the material transferred" but "(e)vidential significance is strengthened by multiple types of transfers" (see "Only Circumstantial Evidence" by Ryland and Houck, in *Mute Witnesses*) and also by location and distribution of the trace evidence on the scene. The importance and significance of all these as well as the long and complicated way to make this understood and valued will be demonstrated and unfolded along the following story.

COLLECTING TRACE EVIDENCE

The scene was examined for evidence of entry or exit. A tiny smear of blood was found on an emergency exit at the far end of the building on the outer surface of the door (Figure 7.2). This was the only indication that someone might have left the building this way after the crime. The emergency exit was the only unlocked door after 8:00 p.m. and also might have offered the opportunity for an unnoticed entrance. It was taken into account that the culprit or one of the

Figure 7.2
(a) Exit. (b) Blood smear on the exit door.

culprits could have hurt himself during the crime. A DNA analysis seemed to be the appropriate means to give more information.

Both victims were stabbed. The male victim had numerous punctures and cuts in his back and chest, and his throat was cut from ear to ear. The female victim had even more stabs and cuts in her body and most of the punctures were located on her back. The victims were found on the kitchen floor, fully clothed with their white uniforms, and both tied to a chair; the victims were also gagged. Their arms and legs were bound and the woman had obviously struggled hard to get her left leg loose.

Initially, a 1:1 taping of the bodies was carried out all over the bodies where the surface was not too wet from the blood or urine, which would affect the taping efficiency. For the tape lifts, 3-in. wide rolls of Scotch Book Tape "845" were used. This kind of tape needs to be ordered in so-called "Kodak quality" to make sure that you will get a proper tape with a clear backing and a clear adhesive without any air pockets. Other products are available on the market that will work but their quality must be assessed by qualified trace evidence professionals before they are used. A metal roller device was used to make the work easier and to prevent contamination. The tapes were numbered and documented by photography (Figure 7.3). As each tape was collected, the tape lifts then were placed on the inner side of a non-adhesive clear bag. To prevent any contamination, the bags were only opened on the scene just before the tapes were placed inside. This kind of tape lifting is a comprehensive method to collect fibers from clothed and unclothed surfaces and offers a possibility to exactly reconstruct the location of each single fiber.

As mentioned before, there were no other unusual items discovered in the vicinity of the victims. Once the taping was complete, the bodies were untied from the chairs and transported to the morgue. A search for fingerprints was

Figure 7.3
1:1 taping e.g. on the back and the legs of the female victim in front.

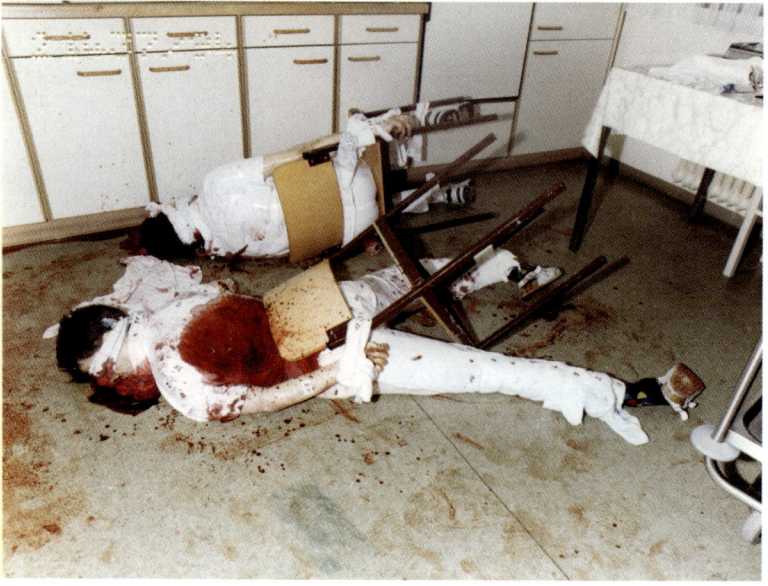

then carried out in the kitchen. The crime scene team learned that the victims changed into their uniforms for the night shift at the hospital. The lockers of both victims which held all their personal belongings were taped by different crime scene personnel (to avoid contamination) to get more information about the fiber "background environment" of both victims. The lockers acted as a kind of archive and contained valuable information because they accumulated fibers from every piece of clothing the victims brought with them to work. These fibers were not only from those items of clothing – non-native fibers from the surfaces of the clothing was also transferred to the lockers. All these fiber types could have been transferred to the uniforms of the victims and also could have been exchanged between them. Establishing this background environment is critical to the success of a trace evidence examination. Finally, to get complete reference materials, different areas of the hospital were also taped; these, however, did not yield much in the way of a textile background.

The victim's clothes and the towels (which had been torn in halves during the crime) from the hospital kitchen were dried, bagged, and sent to the crime laboratory for further investigation and collection of reference material. Because of the apparent lack of evidence and the resulting importance this placed on the fiber evidence, the clothing was taped again at the laboratory. This was not a 1:1 taping because it was expected that the fibers on the surface of the clothing would have moved from their original location during the evidence collection process. The tapings were localized, however, to retain as much location information as possible. The garments were taped inside and out, to capture any differences in fiber population.

AN INTERROGATION WITH NO WITNESSES

There were no direct witnesses regarding this case but nevertheless indications of a previous argument, possibly concerning children, had been heard. The families of both victims immediately came into the focus of the police investigation. The female victim had led a harmonious life and had no enemies. She had planned to go on vacation in the next couple of days with her fiancé and they intended to get married shortly after their return.

The male victim lived with his mother and was divorced. He had two small children who lived with his former wife and his former friend. Both he and his ex-wife were in separate relationships but lived together although both of them had their own apartment. The ex-friend of the male victim was also determined to take over the role as father of the children. The victim was not allowed to see his children by his former wife and his former friend but the day before the murder a magistrates' court ruled that he would get permission to visit them. A visit had been scheduled for the next day, April 14th.

Family members of the victims and close associates of the families were interviewed in hope that this might give an indication of a motive. It was learned that the male victim's former friend had argued with the victim before and also had threatened the male victim. This drew more attention from the police and the investigations focused more on this former so-called friend. On his first interview, the former friend's clothing was taped for fibers to get some information of his fiber background environment.

COURSE OF FIBER INVESTIGATION

In the laboratory, the tapes from the crime scene were checked with a low power stereomicroscope at low magnification. Later, different techniques for further comparison were used to investigate if any of the fibers were consistent with each other or a known source. Besides low power microscopy, a high power microscope with polarized light and fluorescence capabilities was also used. Solubility tests and FTIR were used to confirm the polymer class and sub-class of the fibers. Color was analyzed with a Zeiss microspectrophotometer in the UV/Vis range (250–760 nm). The white cotton hospital uniforms made it easy to classify and segregate the crime scene fibers; furthermore, the fiber background environment in the surrounding of the victims was not colorful (drab hospital colors) and it was a very easy background to search. Laboratory comparison of the tapes from the crime scene very quickly revealed that both victims had an enormous number of black rayon fibers on their clothes and also on the surface of unclad areas (like their arms). The tape lifts from the inner side of the textiles did not show any of these fibers. The rayon fibers appeared black or black fading to colorless in small areas (Figure 7.4).

Subsequent examination of a large representative number of fibers from this population followed and all the fibers were consistent, meaning that the fibers are indistinguishable in chemistry, morphological features, diameter, color and the manner in which the color was applied to the fibers. The appearance of the fibers in this case indicated that the dye was applied to the product later on during the production of the textile – the dye could have been applied to the yarns (yarn dyeing) or more likely to the fabric (piece dyeing).

The relevance of the recovered trace evidence now had to be assessed. The source of the black rayon fibers had to be found – if they could be found anywhere in the hospital, for example, the victims could have acquired the fibers in this area. Additionally, the victims' living areas had to be searched to determine if the fibers came from those areas. If so, at least one victim could have transferred these fibers to the other victim through a close contact, although this did not seem very likely. To investigate the fiber background of the living areas of both the victims, the tapes and reference materials from the lockers

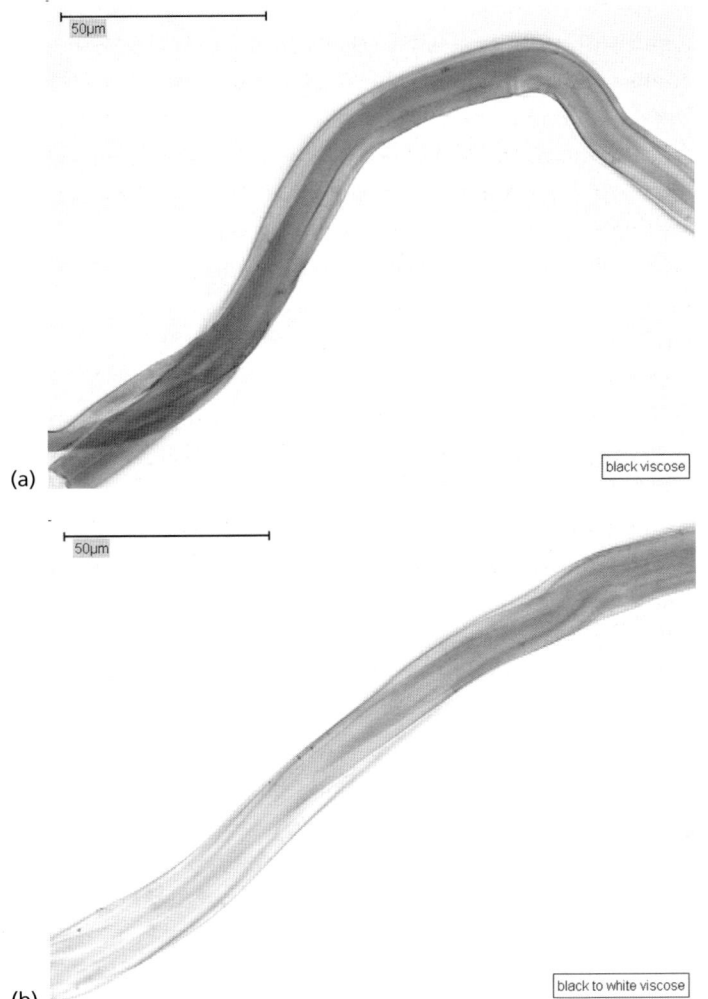

Figure 7.4
(a) Black viscose. (b) Black viscose fading to white.

were compared. The victims changed into their hospital uniforms and left the private clothing into these lockers; as a result, the lockers accumulated a lot of fibers from the textile background of the victims, from the original clothes (direct fibers) and the fibers they brought in on the surface of those textiles (indirect fibers). The tapes from the lockers were compared with the black rayon fibers. Ultimately, the source of the rayon fibers was neither in the hospital surroundings nor in the victims' lockers.

To find such a large number of fibers on the victims' bodies (thousands of the rayon fibers were found on clothed and unclothed surfaces), there must have been a very close and intense contact with a high fiber-shedding textile. This kind of transfer could have happened just shortly before or during the crime, or possibly in the time between the actual murder and the taping procedure. A transfer shortly before the crime seemed very unlikely because the black rayon fibers were concentrated in areas where the culprit must have been in very close and intense contact with the victims – the arms and legs, which were bound to the chairs. On the female victim, a concentration of fibers was found on her right front shoulder, left thigh, the left hollow of the knee, and right lower leg. In total, several thousand fibers were on the tapes taken on the crime scene. On the male victim, the highest fiber concentrations were found on the blindfold, his right upper arm, right back of the hand, and around his right shoulder blade. In total, the male victim had nearly 2000 fibers of the same populations present. Additional tapes which were taken from the clothes of both victims in the laboratory increased the number of evidentiary fibers.

The difference in fiber populations on the outer and inner surface of the garments also suggested that this fiber transfer could not have resulted from long-term or repeated contact in the hospital surroundings. A long-term exposure necessarily would have led to further fiber distribution and, therefore, fibers should be found on the outside of a garment and on the inside as well. The relevance of these fiber traces seemed to be beyond doubt.

The tapes from the ex-friend's clothing, taken from his scarf, his jacket (inside, outside, and the pockets), his T-shirt, and trousers were searched for black rayon fibers. About 120 black rayon fibers were found and they were consistent with the fibers found on both victims. The fibers on both victims and on the clothes of the male victim's former friend showed the same range of variation, including the color change. The recovered fibers from the ex-friend indicated that the interviewed person must have been in recent contact with a textile which shed the same type of fibers as were found on the bodies. This did not mean necessarily that the ex-friend owned the textile but there was a need to get more information from the living area of this person, who now had became a suspect.

COLLECTING REFERENCE MATERIAL

It was necessary to check the fiber environment of the suspect's living areas. This meant necessarily not only his own apartment but also the apartment of his live-in girl friend. There was also another aspect to be taken into account: a friend of the suspect sometimes stayed in the suspect's apartment but did not use the apartment on a regular basis. He also claimed that he never used the suspect's closet. The textiles of the suspect's friend were therefore also of some interest for further investigation.

Fifty-seven background tapes were taken from the suspect's apartment as well as 80 tapes from the apartment of his girlfriend. In addition, 22 tapes were taken from the textiles of the part-time inhabitant of the suspect's apartment. The intention was to get a representative cross-section of the fiber populations in all living areas and on the clothes of the part-time inhabitant of the suspect's apartment. The intention also was to get an overview of the fiber populations which were prominent in the living areas of these people at the time of the crime and if the fibers could be found in areas which provided open access for visitors (like sofas in the living room) or which were in general closed to the public (like closets or storerooms).

ANALYSIS

The first examination of the suspect's background tapes quickly revealed that a certain type of black rayon fibers was present in the suspect's apartment and also in the apartment of the suspect's girlfriend – both apartments were heavily populated with these fibers. The first person, who also came to mind in this connection, was the suspect's friend who occasionally stayed at the apartment. The tapes which were taken from his clothes were searched to learn if the source for all the fibers could be found in his belongings but only several of the black rayon fibers here and there on the surface of some textiles were found. This part-time resident was in contact with the apartment of the suspect and a few single fibers of a textile which shed heavily all over the place were expected. No significant number of fibers were found which was worthwhile for further investigation. The original shedding textile could not have been among the belongings of the part-time resident. Likewise, he also claimed that he kept his clothes separate from the resident's clothes and stored them in his own bags.

The close investigation of all the background tapes from both apartments revealed that there were fibers all over both places which were consistent in all tested features to the black rayon fibers. The fibers were found in places which were open to public access, such as seats and the bed, and also in places which are generally not open to "public" access, such as cupboards, closets, store rooms, and an ironing board in the store room. This indicated that there must have been a textile around which was present in this living area for a long period of time and had the chance and shedding ability to transfer fibers to all of those places. But the source of the black rayon fibers could not be found in any of the apartments.

According to all the blood on the crime scene it seemed very likely that the culprit's clothes must have been contaminated with the blood of both victims and that the textiles were hidden or just thrown away. The victims were highly contaminated with black and black fading to colorless viscose fibers. To prove the thesis right that those fibers were relevant, there had to be other direct or indirect fiber transfers from the living area of the suspect.

A direct fiber transfer is the transfer of a textile's own fibers to another surface. An indirect transfer means that the fibers deposited on one surface are then carried by that source to another location and deposited there. Other directly transferred fibers could be expected in other locations than the crime scene if the transfer resulted from different textiles worn on different parts of the body (such as shirts, trousers, or socks). Directly transferred fibers could also be expected nearby if the transfer resulted from one textile with different fiber types in different parts of the whole pattern (e.g. waistband, collar, or cuffs) or from different textiles worn together (like, a jacket and shirt cuffs or a jacket and sweater).

Indirectly transferred fibers will always be present and can be found particularly in all the areas highly populated with directly transferred fibers which are transferred through a direct contact. Therefore, all of the areas with numerous black rayon fibers were searched for another fiber type and a population of brown-black cotton fibers was found. These brown-black cotton fibers were always related to areas with a large number of the black rayon fibers, including tapes from the two closets of the suspect's apartment (Figure 7.5). These closets contained jackets and shirts of the suspect but the ultimate source of either fiber type could not be found.

As the crime scene shows, the culprit must have been in contact with nearly every part of the victims' bodies and especially to the bound extremities. Contact with the sleeves of an upper garment seemed to be very likely and fit in the context of the crime circumstances. The large number of black-brown cotton fibers indicated that most likely they were transferred through a direct contact and that the textile must have been worn in very close relationship to the one which shed all of the black rayon fibers. It is important to remember that both fiber types would not necessarily originate from the same garment.

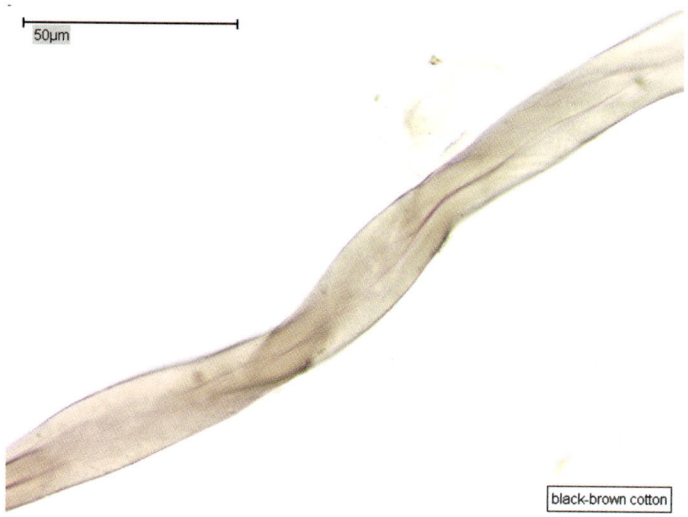

Figure 7.5
Black-brown cotton.

The distribution of the black viscose fibers and the brown-black cotton fibers led to the same conclusion. On the one hand, both fiber types were present in one of the closets but, on the other hand, the black rayon fibers appeared on other tapes from many items in both apartments but without the black-brown cotton type. The source of the brown-black cotton fibers must have been in contact with the same area as the black rayon fibers and therefore the closets came into focus for further investigation. On one of the shelves, gray-blue cotton fibers, black polyester fibers and gray acrylic fibers were found which were consistent in their morphological features, chemistry, color, and method of dye application with other fibers found on both victims (Figure 7.6). All of the tapes from the apartment of the suspect's girl friend, from his co-resident's belongings, and from different people the suspect was in recent contact with were examined, but these fibers were found only in one of the closets of the suspect's apartment.

Why were those three fiber types (the blue cotton, black polyester, and gray acrylic fibers) out of all the other fibers? A tiny fiber "pill" was found on one of the tapes from a binding around the male victim's left wrist. This pilling contained many types of fibers but only those three different types were consistent with fibers from the suspect's closet. Pills form on the surface of a textile from normal wear and contain the "native" fibers of the originating textile source. In general, these pills are fixed to the textile surface of the source by a few so-called "anchor" fibers. In addition, fibers from other textiles ("non-native") may build up around those anchor fibers and create a complex of different fibers. Other than the anchor fibers from the originating textile source, these pills do not necessarily contain the same fiber populations. It is more likely that different pills from the same garment also contain different types of fibers. Therefore, it was a combination of extraordinary good luck and, most likely, repeated contact between related textiles to find pills with three consistent fiber types. On this basis further examinations were carried out and single fibers of those three fiber types were found on both victims.

Another course of investigation was also followed. A number of the black rayon fibers were found on the outside and inside of the clothes, specifically a jacket, the suspect wore on his first interview. The large number of fibers indicated that the black textile must have been in recent contact with the other sources. Therefore, fibers from other sources could have been transferred to the black textile and also could have been transferred to the victims through an indirect transfer. A close examination of the tapes from the crime scene showed that single fibers consistent with two fiber components (red polyester, dark gray polyester) from the jacket could be found on both victims (Figures 7.7 and 7.8).

Laboratory examination of other physical evidence in the case revealed only DNA from the female victim at the exit of the building but this DNA was mixed with another profile which did not match the DNA of the suspect. No

Figure 7.6
(a) Gray-blue cotton (also component of the little fiber pilling). (b) Black polyester (also component of the little fiber pilling). (c) Gray acrylics (also component of the little fiber pilling).

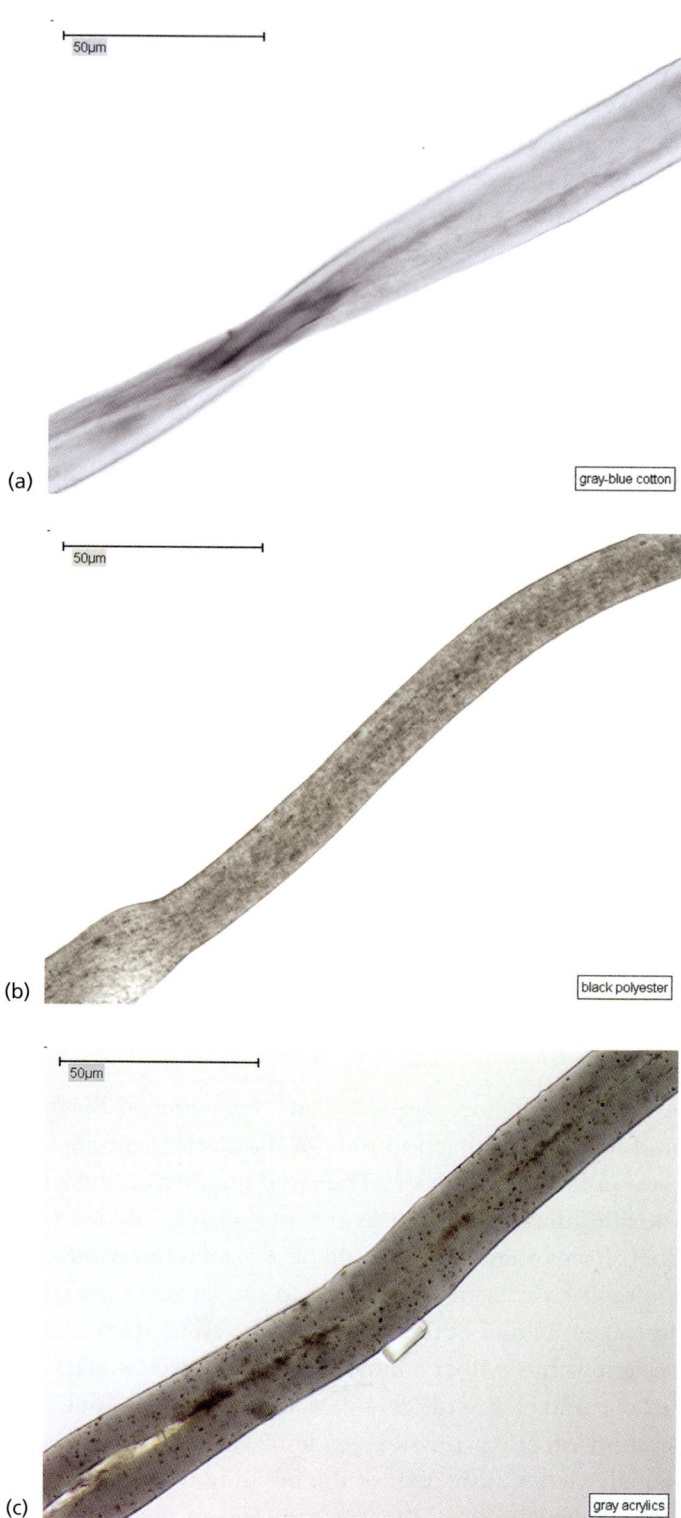

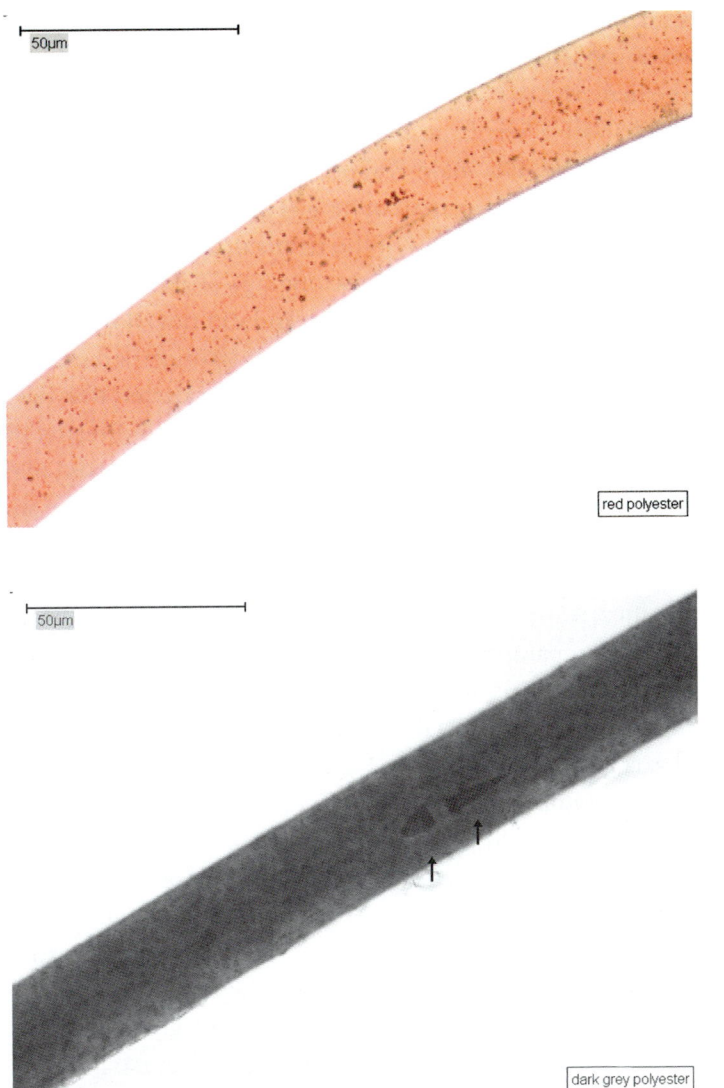

Figure 7.7
Red polyester.

Figure 7.8
Dark gray polyester.

case-related person matching this DNA was identified. A closer look at Figure 7.2b reveals that the tiny bloodstain was found on an area of the door which also showed smears and fingerprints which could have resulted from a long-term use of the door.

This seemed to be a very good result with thousands of black rayon fibers on both victims obviously resulting from an intense direct contact. There was a second complex of fibers (black-brown cotton), which was also most likely directly transferred to the victims' bodies and belonged to a textile the culprit must have worn at the scene. Five more indirectly transferred fiber components (gray-blue cotton, black polyester, gray acrylic, red polyester, and dark gray polyester) each with a different number of fibers were found on the tapings from the crime scene. In addition, a fiber pill was found which was located on a binding around the male victim's left wrist. All of the recovered fibers yielded the same analytical results with fibers from the living area of the suspect (the former friend of the male victim). The main component was found in the apartment of the suspect and also in the apartment of his girl friend. All of the other fibers were only connected to the apartment of the suspect (closet) or to the clothes he wore on his first interview (jacket). The belongings of the suspect's co-resident were not a source of these fiber components and did not show a relationship to the crime.

The prosecuting attorney was not very concerned over a solely "circumstantial evidence" case. She considered it could be argued that the male victim did have opportunity to be in contact (either directly or indirectly) with the suspect but in fact the large number of transferred fibers and the even higher number of fibers on the female victim which had no relationship with the suspect worked against such a hypothesis. As Ryland and Houck pointed out, "(i)f the individuals involved in the alleged transfer are known to have been in contact with one another prior to the incident in question, any transfer discovered typically becomes meaningless unless addressing a reconstruction of events." The reconstruction of events was partly possible due to the 1:1 taping. The advantage of 1:1 taping that one can establish a direct link to the crime scene via fiber distribution charts and fiber mapping (Figures 7.9 and 7.10). A very good example can be seen by comparing Figures 7.1 and 7.9. The bloody imprint on the back of the male victim could not be interpreted but the mark must have been produced during the crime took place and an amount of blood was already present on the crime scene. The fiber map also showed high fiber transfers area exactly in the same region below the right shoulder blade of the male victim. The transfer area showed the well-known black rayon fibers. Without the 1:1 taping, there never would have been a chance to locate this spot and connect the fibers to the crime and the actions which took place during the actual murder.

Direct and indirect fiber contacts can be visualized by a fiber distribution scheme if a 1:1 taping is carried out. The relations between the actions on the scene and the fiber-contaminated areas can clearly be seen. For example, both victims were tied to the chairs and therefore contact between crime-related areas, like the lower legs of the female victim, must have taken place. The culprit must have restrained the legs somehow to get the ankles tied to the chair because the woman struggled hard to get her left leg loose – the imprints in the blood on the kitchen floor which originate from her shoes confirmed this. The high transfer of fibers to the female victim's lower leg was plausible in this respect and therefore proved to be crime related.

From a fibers analyst's point of view, it was one of the most convincing cases to prove a contact without the source garments of the transferred fiber populations

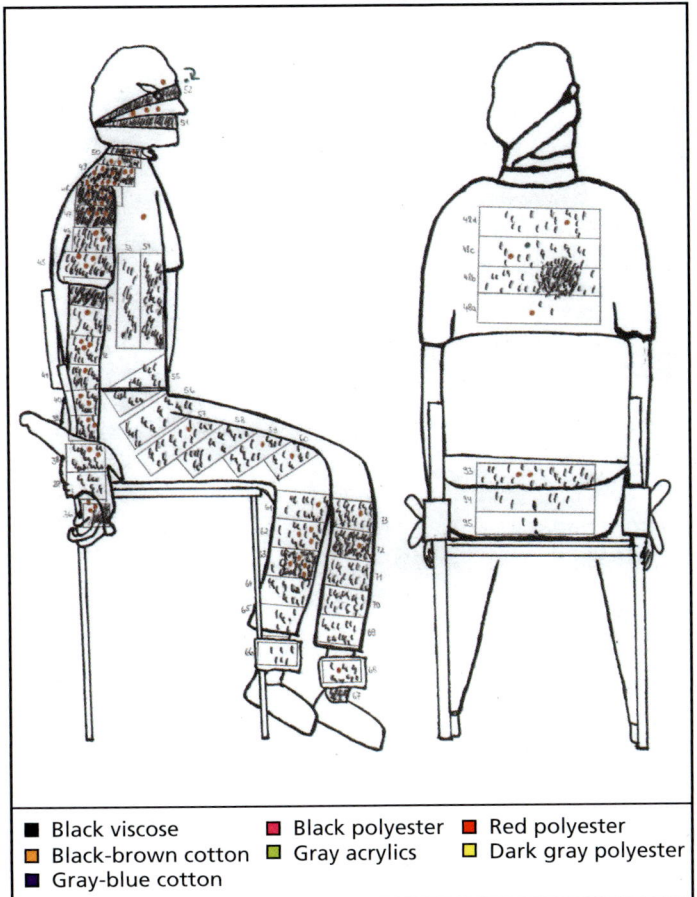

Figure 7.9
Fiber distribution scheme of the male victim.

present. Everything seemed very clear and the case went to the district court with seven types of indistinguishable fiber components connecting the populations found on both victims and in the suspect's living area, especially in the suspect's own apartment and in places not open to the public. It was important to state the relevance of the fibers transferred on the victims and present the complex network of fiber relationships (Figure 7.11) (see also Neubert-Kirfel 6/2000).

The fiber testimony was lengthy and lasted a couple of days. Although, or perhaps because, every single fiber type was discussed in great detail, the complex network of fiber relationships got out of focus. The court wanted to discuss every single fiber type and its special relevance from the point of the individual fiber but, as forensic scientists all know, a part of something is not the whole. Try to imagine that someone cuts a painting into small differently colored pieces, only collects the blue ones, and asks you to tell if the blue snippets belong to a Picasso or a Matisse. Without the composition and complexity of the colors, all the information that would indicate Picasso is gone (as Neubert-Kirfel pointed out again

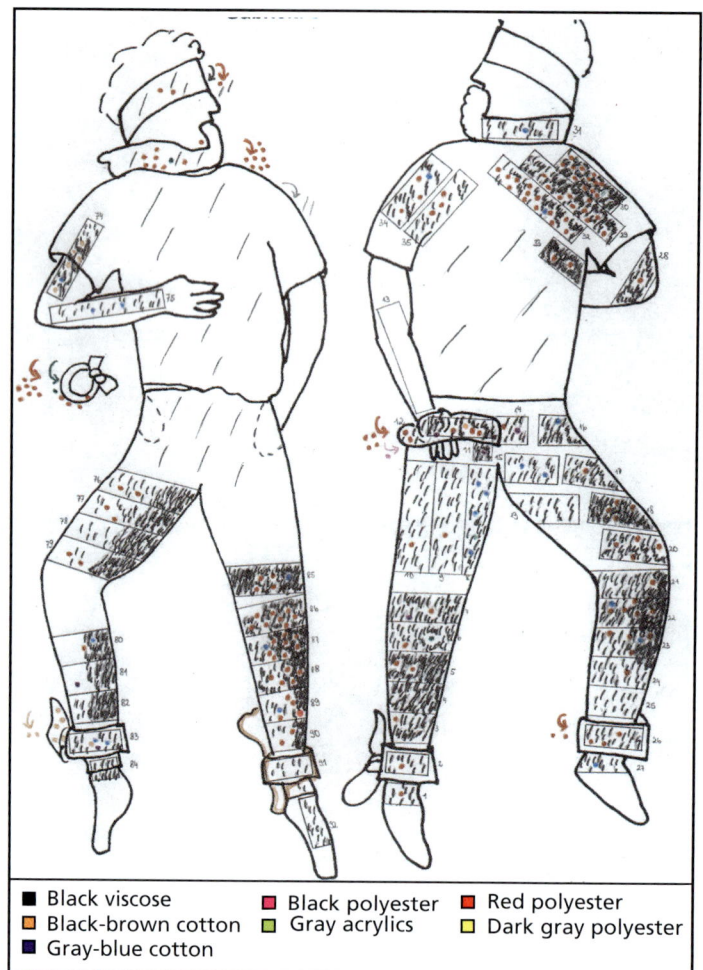

Figure 7.10
Fiber distribution scheme of the female victim.

on a meeting in 1995). The more questions that were asked about a single fiber type or component, the more the whole case became confused and unfocussed.

Finally, the court released the defendant with an acquittal and stated part of the reason was that some of the recovered DNA did not belong to the victim or the suspect. The court reasoned that the culprit must have injured himself during the stabbing to leave his DNA on the door and, therefore, the suspect could not be the one who murdered both victims. In another part of the judgment, the court reasoned that the culprit must have worn gloves because no fingerprints were found on the scene! Under these circumstances, the court felt that the fiber evidence was not strong enough to convict the suspect.

The prosecuting attorney was not satisfied with this result and she filed an objection. Under German law, an objection can be filed for to get a case to trial again. The first step is to state the grounds of appeal as an objective reprimand

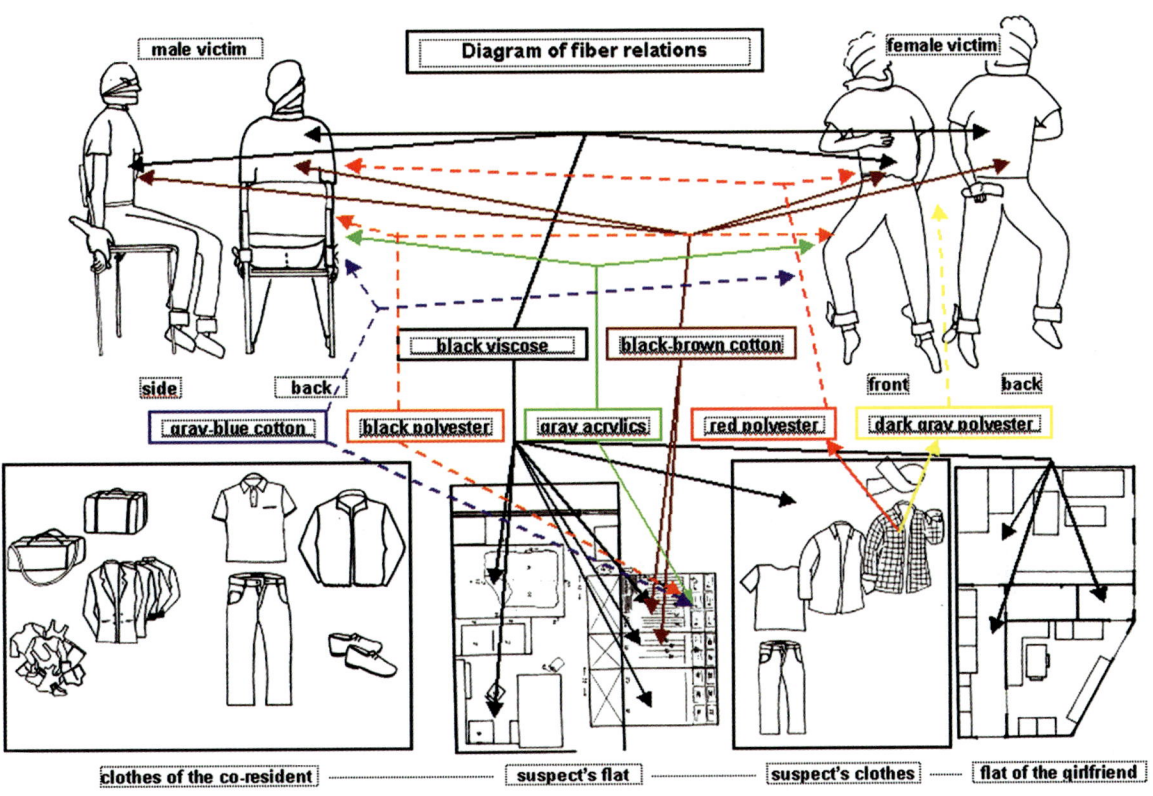

Figure 7.11
Diagram of fiber relations.

and get the motion to the Supreme Court (Court of Appeal). The prosecuting attorney claimed that the State Court violated objective rights, stating that the State Court did not follow the fiber expert's evaluation of results and failed to explain in the judgment on what grounds the court choose to do so. She also claimed that the court ignored former fiber evidence-based judgments of other State Courts and also of the Federal German Supreme Court which is the highest German Court. The Court of Appeal granted a review on behalf of the prosecuting attorney and was to be carried out by the Federal German Supreme Court. The case would now be represented by the Federal Public Prosecutor at the Federal German Supreme Court. Based on the review, the Federal German Supreme Court ruled that there were objective reasons and that the case should go to trial again at a different State Court.

The Federal German Supreme Court found that the first State Court failed to evaluate the DNA and the fiber results adequately. The Federal German Supreme Court stated also that the DNA result should be valued as a neutral result because the source of the unknown DNA is not necessarily related to the crime (the DNA could have been left behind by a completely uninvolved person – remember the fingerprints and smears on the door). The Federal Court reprimanded the State Court and reasoned that the State Court failed to acknowledge the evidential value of the fiber results as a whole:

> "... The value of a textile fiber result depends on the question how probable it is that a fiber combination can be found by chance on any textile which is not related to the crime ... The court has to evaluate if the fiber results represent a characteristic image ... and if the image is that special that the presence of an identical fiber combination in the living area of a completely uninvolved person is a coincidence which is far from the thoughts of somebody ... The court has to evaluate if the found fiber combinations represent a characteristic fiber result. Under these circumstances the number of found fibers is essential. The presence of cross transferred fibers can be also essential ... The State Court didn't value the fiber results as a whole." (This judgement was based e.g. on Adolf, 1990.)

The State Court discussed the value of each component of evidence. More than one year after the first trial, the case went to court a second time. In addition to all results from the original casework, the court wanted additional information. The laboratory was instructed to collect and document the appearance of a variety of fibers to evaluate how alike or how different such a range of fibers is. Fibers of all the fiber types relevant in the case were collected to relate this documentation of fiber components to the case. Friends, family, and co-workers were asked to check their personal textile belongings and bring textiles from their environment to represent the seven components of the case (black rayon, black-brown cotton, gray-blue cotton, black polyester, gray acrylics, red polyester, and dark gray polyester). To supplement what people brought in to include all seven fiber components (the fact that no one had all of the same fiber components in their textiles should have told the court something important about fiber evidence!), some fabrics were purchased at a department store. In total, 151 samples were collected.

None of the textile environments sampled provided all seven relevant fiber types. Moreover, every single fiber type could be distinguished from all the other fibers of the same general group. All the fibers of one type and color showed a wide variety in shade (UV/Vis) and other characteristics. For example, the polyester fibers were very similar in their cross-sectional shapes but varied largely in diameter and delustering, while the acrylic fibers varied more in cross-sectional shape, diameter, and color shade. The examination of the fibers of the same type but from different garments also showed variance resulting from production, such as stretching marks and chemical finishes on the surface.

The court discussed every single fiber type and became aware that the fibers of one single type showed a wide range of variations which depends on the fiber type, color, shade, application of dye, and side effects resulting from production. The court also determined that fibers originating from the same garment can show small differences in appearance but that the variance is limited and this bounded variance must be reflected in the recovered fibers. The court also considered the demonstrated relationships between the fibers and visualized the connection between the crime and the fiber distribution (e.g. see Figures 7.1, 7.9, and 7.10). The two direct fiber transfers and five indirect fiber transfers led only to the suspect and only into his own apartment and closet. Neither the girlfriend's apartment nor the textile belongings of the suspect's part-time co-resident had the same relevant fiber combinations. Therefore, this court evaluated the chance of an identical fiber combination in the living area of a completely uninvolved person as an improbable coincidence.

The DNA evidence from the exit door of the hospital was also taken into account and the DNA expert stated that the DNA could have been present on the door prior to the crime and the DNA could have also resulted from a biological fluid other than blood.

SUMMARY

After the first fiber transfer was discovered, the 1:1 taping helped to relate this fiber to the crime by mapping these to the other different fiber transfers. This first complex was a valuable investigative lead for further comparison and led to the living area of the suspect.*

The comparison between the fibers recovered from the victims' bodies and the living area of the suspect indicated that it was not unreasonable to assume that the source textile shed fibers easily. The combination of seven different fiber types from the apartment and closet of the suspect strongly indicated the suspect was in contact with both victims.

A study of the fibers collected from colleagues and purchased at a retail store points out that people own very independent and individual collections of clothes and textiles. It is difficult to collect seven different types of fibers in a certain color from only one person. In addition, it became clear that fibers made from the same polymer can be very different in appearance. Fibers were differentiated either by variations in diameter, morphological features, cross-sectional shape, color, shade, and dye application. Other differing characteristics are imparted to the fibers due to the varieties in production. It is very unlikely that different fiber types most likely produced from different manufacturers would correspond coincidentally in all of these independent characteristics. All of the characteristics of each single fiber type from the seven would have to have coincided simultaneously. A single fiber may not mean much, but a combination of different and multiply transferred fiber types strengthens the results enormously (see also Biermann and Grieve, 5/2001).

This investigation served to illustrate the value of fiber evidence in a "circumstantial" evidence case. It also revealed that a poor interpretation of DNA evidence can lead to wrong conclusions. The DNA evidence from the exit door

*See Michaud's chapter in *Mute Witnesses* for another example of trace evidence acting as lead value.

of the hospital did not affect the final judgment because the unidentified DNA could have been at the door before the crime and the presence of another DNA profile did not lead inevitably to the assumption that it was the "real" culprit's. The potential complications in the interpretation of trace evidence also were obvious during this investigation: a known or possible prior contact between both victims and the suspect clearly would have affected the investigation. Trace evidence is everywhere but their context in relation to the crime is what is most important (Decke, 7/2000).

The suspect refused to give evidence and never commented on the results. In the second trial, the State Court found him guilty and sentenced him to life without the possibility of parole after 15 years.

ACKNOWLEDGEMENTS

I would like to thank all my colleagues from the LKA working on this case. Every day, their skills, insight, and questions helped to fill voids and to find new aspects of the case. I would also like to thank Dr. Neubert-Kirfel from the LKA in Nordrhein-Westfalen, Germany, for her full support between the trials. The time, knowledge, and skill of Ms. Neubert-Kirfel in reviewing the documents provided me with a lot of background information and is much appreciated. Also, a word of appreciation to Ms. S. Nielsen, Attorney for the Judicial Circuit for the State of Berlin, Germany, for her effort in this trial. Her work was well rewarded by enabling her to provide effective "translation" between the scientists and the courts. All of the court officials tried very hard and made a genuine effort to understand the science. Special thanks also to S. Nehse for reviewing the manuscript.

REFERENCES

Adolf, F.-P. (1990) NStZ, 65, 70.

Biermann, T. and Grieve, M. (5/2001) *Kriminalistik*, 337–340.

Decke, U. (7/2000) *Kriminalistik*, 467–472.

Locard, E. (1928) *Police Journal*, 1(2), 177–192.

Locard, E. (1930) *American Journal of Police Sciences*, 1(3), 276–298; 1(4), 401–418; 1(5), 496–514.

Neubert-Kirfel, D. (6/2000) *Kriminalistik*, 398–404.

Ryland, S. and Houck, M. (2000) "Only circumstantial evidence," in *Mute Witnesses*. New York: Academic Press.

CHAPTER 8

WHO DO YOU BELIEVE?

Barbara P. Wheeler
Forensic Scientist, Lexington, KY
(Former Forensic Laboratory Supervisor, Trace Evidence Unit
Kentucky State Police Forensic Laboratory Section)

INTRODUCTION

The month of September in 1993 brought with it a circumstance that would drastically change the lives of many individuals within two Kentucky families. Two men – Jim and Tommy – though not blood-related (Jim's brother-in-law was Tommy's cousin), but who commonly referred to each other as first cousins, were the best of friends and almost constant companions. Jim would later state, "No one ever saw him without me".

Early in the evening of September 26th, Jim and Tommy left Jim's family-owned hotel in his grandfather's 1988 Mercedes 560 SL convertible. Embarking on what they thought would be a fun evening of drinking and partying, they went to a popular local bar. To continue their partying, around midnight, they decided to proceed to an after hours club. As they drove towards the club, Jim voiced his desire to freshen up and change clothes.

The convertible made its way through the college-town streets towards the apartment. Reaching a wide, two-lane roadway, they proceeded at a high rate of speed. At some point, the driver lost control of the vehicle and left the road, eventually striking a brick wall. Since neither Jim nor Tommy was restrained, both were ejected. Witnesses quickly alerted the local police department who responded along with the emergency services. Both Jim and Tommy were quickly stabilized and transported to a local hospital within a few miles drive of the accident scene. The witnesses could not provide many details of the accident to the officers; all they could account to was bodies flying out of the vehicle. According to the officer, "there was no way to tell who was who, because they were the same age, same size and were wearing similar clothes".

It is important that all those involved in forensic investigations realize the importance of securing all potential evidence from a scene so that the science can provide the truth. Many times, unfortunately, the trace evidence is left at a scene or in a police department evidence room and never analyzed. Because of this fact, many unanswered questions, thus, remain unanswered. Trace evidence is overlooked for a variety of reasons: commonly, it is a lack of knowledge as to the potential of trace evidence. The value of trace evidence becomes apparent with each examination. A good trace analyst can examine and analyze almost anything. Even if the value of an item is unknown at the time of the crime, an officer should collect all possible evidence items, as he may not have the opportunity to go back and retrieve the item at a later time. Any known standards, which may be needed for comparison, should also be collected. Ensuring that all possible evidence is available for analysis is only the first step. It is just as important for all those involved in the investigation to be aware of this "collection of items", so that all items are analyzed if it becomes necessary.

CRIME SCENE

The accident scene was confined to a short stretch of upward roadway with a very slight right-hand curve. The roadway had a speed limit of 45 mph, and was a wide, two-lane road with a large, grassy area on one side, and a sidewalk accompanied by a brick wall on the other. The long expanse of brick wall was broken-up by pillar insets at regular intervals.

Upon arrival at the scene, the officer noted major damage to the brick wall, and a single, overturned vehicle partially in and off the roadway into the grassy area. Two individuals had been ejected from the vehicle.

From the initial examination, it appeared that the Mercedes had been traveling at a high rate of speed, striking the brick wall in numerous locations. Both occupants had been ejected at some point. Vehicle tire marks, various fluids, and vehicle parts were scattered along the area. Tommy was lying in a prone position on his right side, to the right of the overturned vehicle. Jim was lying to the left of the vehicle, on the sidewalk. Both individuals were initially unresponsive and quickly transported to the hospital.

All impact areas to the vehicle and scene were documented in the officer's notes. Various loose vehicle parts, clothing items, and the vehicle itself were collected from the scene. Since this seemed to be a single-vehicle injury accident which could be followed up on later, the scene was quickly cleared. Alcohol was listed as a possible factor in the accident (see Figures 8.1 and 8.2).

After their arrival at the hospital, both individuals were examined. The doctors determined that the injuries to both Jim and Tommy were more extensive than originally thought by those attending the individuals at the accident scene. Jim appeared to have a severe head injury with extensive abrasions to the chest and back areas. He also complained of extreme pain in his lower legs and feet. Shortly after arrival at the hospital, Tommy took a turn for the worse and died of what was later to be determined as acute cardiorespiratory insufficiency, due

Figure 8.1
Scene photograph, wall and road.

Figure 8.2
Scene photograph, Mercedes.

to acute peritoneal hemorrhage and blunt force trauma. The simple single-vehicle accident quickly turned into a murder investigation. It now became important to determine everything about the accident itself and even more importantly, who the driver of the vehicle was and what additional factors, if any, played into the accident.

INVESTIGATION

After learning of the death, the officer began a complete accident reconstruction and investigation to determine the path of the vehicle and occupants.

Further examination of the accident scene revealed that the Mercedes, which had been traveling at a high rate of speed, began to yaw prior to striking the wall. Numerous areas of damage to the brick wall were noted, in addition to various vehicle tire marks, fluids, and parts. It appeared that the vehicle, reaching critical speed, yawed in a counter-clockwise direction and left the roadway, crossing the sidewalk and striking the brick wall with the front end of the vehicle. Damage to the brick wall was noted, and skid or tire marks were found in this area. Fluid was spattered along the brick wall in this area. It appeared that the right front fender of the vehicle had impacted with a brick pillar, nearly ripping the front end of the vehicle off. The front bumper was deposited at this spot. The oil pan was also ripped open by this impact, and further fluids were deposited on the brick wall. This major impact accelerated the vehicle's counter-clockwise motion, rotating the front axle 90 degrees. As the vehicle continued to rotate, additional portions of the brick wall were knocked out, causing further damage to the right side of the vehicle. Various transfers and tire marks showed the vehicles' progress. The right rear fender impacted another brick pillar, causing the vehicle to change directions, and send it back towards the road. At that point, while continuing to rotate in a counter-clockwise direction, the vehicle hit the curb, which sent the front end of the vehicle downward. This caused the front end of the vehicle to dig into the grass, causing the vehicle to flip. The vehicle finally came to rest on its roof.

The officer conducted drag-sled tests at the scene to determine the approximate speed at which the vehicle was traveling, when the accident occurred. Using the critical speed formula, he determined that the Mercedes was traveling between 85 and 92 mph just prior to striking the brick wall. The area itself was documented using a total station survey instrument. Physical evidence found at the scene, and the "positions" of the vehicle during the accident, were added to develop a scale drawing of the accident. Through this, he was able to determine the actual path which the vehicle took; the path of the occupants, however, was still to be determined (see Figure 8.3).

From studying the vehicle dynamics and the occupant kinematics, the officer concluded that the air bag would have deployed when the initial impact of the vehicle's front end with the brick wall occurred. Since the driver was unrestrained, although protected from serious chest injuries by the air bag, he would have continued to move forward. This allowed the driver to strike the lower dashboard, steering wheel and steering wheel column. The counter-clockwise movement of the vehicle would continue to keep the driver close to the dashboard. Since the dashboard and steering wheel would be in close proximity, the deployment of the air bag from the steering wheel would force the driver's arms and hands to move upward. This would allow the driver to possibly strike the windshield. The passenger, also unrestrained, would also move forward. Since the movement of the vehicle was counter-clockwise, however,

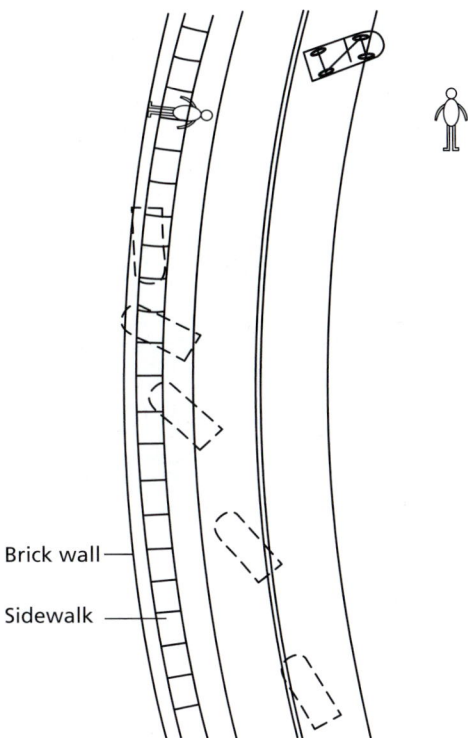

Figure 8.3
Accident scene sketch.

the dashboard and windshield would actually be moving away from the passenger. Any contact for the passenger would have occurred with the passenger side door.

Once this theory was determined from an examination of the vehicle and the occupant kinematics, it now became important to develop actual physical evidence to support the theory. During the initial scene investigation, the officer had collected portions of the Mercedes interior. Head hair samples and eyebrow hair samples had been collected from both Jim and Tommy. In addition to this evidence, part of their clothing items were also forwarded to the Kentucky State Police Forensic Laboratory for analysis in the Trace Evidence Unit. Following the receipt of the laboratory report results, additional clothing and vehicle parts, which had been secured, were also forwarded. After receipt of the second laboratory report results, even more vehicle parts, which had also been secured, were forwarded for comparisons. Through these submissions and examinations, the physical evidence was beginning to mount and making it possible for the officer to place a driver in the driver's seat.

ANALYSIS

Since many of the vehicle dynamics could be easily calculated, the main challenge presented to the officer was the absence of an eyewitness who could place

Figure 8.4
Photograph, dashboard area with speaker grill.

the driver behind the steering wheel at the time of the accident. The physical evidence would have to become the "eyewitness". This challenge is frequently presented to forensic scientists on cases: to put as many pieces of the puzzle back together again as possible, through the use of good science. Using science, the truth can be determined.

Vehicular accidents, by their nature, imply the possibility of violent impact. In the field of accident reconstruction and investigation, officers and forensic scientists take advantage of this fact. Whether the accident was a single, multiple, or pedestrian–vehicle investigation, the possibility of trace evidence transfers is always present. One manner, in which the value of the trace evidence becomes apparent, is when the evidence is analyzed as part of these violent contact investigations. Recognition, collection, and analysis of the trace evidence can be of great benefit to the investigation if processed and interpreted correctly.

The original submission of evidence to the Trace Evidence Unit included only a few pieces from the interior of the Mercedes, the windshield and some clothing from the occupants. One of the pieces submitted was the vehicles' left lower speaker grill, as the officer had noticed areas of contact and thought that a transfer to the pants of either individual might be present (see Figure 8.4). Hair had been collected from the windshield above the steering wheel area, and the windshield itself had also been submitted for analysis. Head hair samples from one individual and eyebrow hair samples from the other individual were submitted. Brake and accelerator pedals, as well as the shoes from both individuals, were also submitted for analysis. Upon initial examination of even these few exhibits, it became clear that with the complete submission of the necessary items, a possible driver could be identified.

*Figure 8.5
Speaker grill.*

The left lower speaker grill was first examined using a stereomicroscope. The speaker grill was composed of a blue thermoplastic with areas of black fabric covering the sound openings. A large area was located on the speaker grill which had been subjected to hard contact with a fibrous item. The blue thermoplastic had been smeared to actually capture and hold numerous fibers from the item, which had come in contact with it (see Figure 8.5). After the fibers were collected, they were examined further and identified as blue cotton and white cotton fibers. Cotton fibers are generally some of the easiest fibers to identify: with a polarized light microscope, cotton, a natural fiber, will appear as a regular-to-irregular-twisted ribbon. The fibers may contain a lumen and convolutions, if the cotton has been mercerized. In most circumstances when a cotton fiber has been finished with this technique, the lumen and any convolutions almost completely disappear. The distinguishing characteristic for cotton fibers is the lack of complete extinction under crossed polars.

Many blue jeans are composed of blue cotton and white cotton, usually woven in a twill pattern. While the dye shading may vary along the fiber and fabric, the composition used for the fibers is very common. Indigo is the most common dye used and is being utilized for most blue jean material produced today. This generally prohibits further associations, since in many situations one pair of blue jeans cannot be distinguished from another. In rare circumstances, the addition

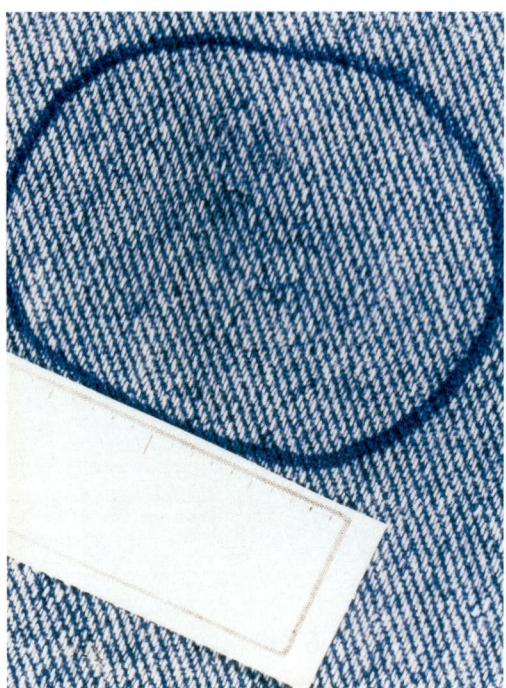

Figure 8.6
Photograph, blue jeans.

of whiteners and brighteners may give the cotton fibers or fabrics a unique quality to help distinguish one pair of blue jeans or denim fabric from another. In this particular case, however, not only did both individuals have on blue jeans, but they were also wearing the same brand. Even though a complete comparison was performed on both pairs of blue jeans, no characteristics could be found to distinguish which pair of blue jeans had left the transfer on the speaker grill. The various shades of coloring of the blue dye in the blue cotton were similar for both pairs of blue jeans, and for the foreign fibers from the speaker grill.

Since the blue jeans themselves would not differentiate a possible source it was surmised that perhaps the speaker grill itself could. Since contact with the speaker grill was sufficient to transfer fibers from the blue jeans onto the speaker grill, the possibility of the speaker grill transferring blue polymeric material to the blue jeans themselves also existed. Both pairs of blue jeans were examined using a stereomicroscope. Both items of clothing had been cut open at the hospital during treatment, and contained many areas with blood and bodily fluids. Upon closer examination for areas of possible contact, only one pair – Jim's – appeared to have a foreign transfer, on the left knee (see Figure 8.6).

There seemed to be a pattern to the transfer, similar to the dimensions of the sound openings in the speaker grill. The overall pattern, though linear in design and consistent with the speaker grill, did not contain enough detail for a positive identification. Closer examination of this transfer area on the other hand, showed numerous smears of what appeared to be blue polymeric material. Analysis and comparisons of this material was performed using Fourier Transform InfraRed (FTIR) spectroscopy and Pyrolysis Gas Chromatography (PGC). FTIR spectroscopy has been used to characterize samples of various types of composition and is valuable in determining the generic class of a polymer. Information on fiber additives or dyes can at times also be distinguished using FTIR spectroscopy. This method of testing is preferred, since it is a quick and non-destructive identification of a sample. Molecules in a sample will absorb infrared radiation at specific frequencies that match the vibrational frequencies of the molecule. FTIR spectroscopy involves the determination of the frequencies at which the absorption occurs, and the preparation of a plot, called an interferogram, of the radiation absorbed versus excitation frequency. This interferogram is then converted to an amplitude–wavenumber spectrum. Identification of a sample can be achieved by comparison to a reference collection or to known standards.

PGC can also be used to characterize of the composition of a sample. This type of testing, however, generally requires more time, and is destructive. PGC may also be used as a comparison technique for some thermoplastics. A polymer will break down when heat is applied; the manner in which these polymers break down and fragment can sometimes be associated with a specific generic polymer class. In forensics, both methods of analysis – FTIR spectroscopy and PGC – are used for a variety of different types of evidence. Commonly in trace evidence examinations, they are utilized for polymer comparisons, such as paint, fibers, rubbers, or thermoplastics. The FTIR spectroscopy and PGC analysis and comparison of the speaker grill thermoplastic, and the foreign polymeric material recovered from the blue jeans, showed that the blue thermoplastic from the speaker grill and the blue polymeric transfers onto the blue jeans were of the same color and general polymeric type. Material had, therefore, been transferred to and from both possible sources. The first clue as to who was driving had been found.

The second piece of evidence to consider, the windshield of the Mercedes, had been submitted along with hair that had been removed from the windshield. This hair had been located above the steering wheel. The initial examination of the windshield showed several areas of breakage; however, some of these areas could have been the result of the vehicle's flipping or could have occurred during the removal of the windshield. It, therefore, became important to determine which breakage points were actual impact areas that occurred prior to the removal of any evidence.

A vehicle's windshield is actually composed of two separate sheets of glass, which are held together by a piece of lamination. Each piece of glass from a windshield should be examined independently. Sheets of glass, whether from a window, mirror, or windshield for instance, are actually super-cooled liquids. When the glass comes in contact with a force or impact, the glass will actually stretch and bend. When the force exerted on the glass is greater than what the glass can absorb through stretching and bending, it will begin to break and fracture. Depending on the type of glass, the method of fracturing or breaking may differ. For most glass, initial fractures radiate outward from the point of impact. Subsequent concentric fractures develop around the point of impact, creating a "spider-web" pattern. Safety glass is treated during production, which places stress in the glass. This causes safety glass to break into small honeycomb-shaped pieces, which is different than other types of glass. Three "spider-web" patterns were located in the windshield removed from the Mercedes, which could be associated with actual contact. The main area of impact was located above the steering wheel and was created by a force applied from the interior of the vehicle. Many of the radial and concentric fractures in this area contained glass with exposed, rough, and jagged edges, which is ideal for the transfer and retention of trace evidence. Upon closer examination, additional hair fragments were collected from these areas. These fragments, along with those which had been collected by the officer, were examined using a compound light microscope. Although the hair fragments were determined to be human, and other microscopic characteristics were visible, a possible body part could not be determined. This precluded any further comparisons with the standards that had been submitted. It should be noted that these hair fragments were extremely short, most likely prohibiting even mitochondrial DNA analysis. A small amount of possible tissue was also collected from the area above the steering wheel, but testing for the possible presence of blood yielded a negative result. The amount of this tissue sample was extremely small, so further biological testing was precluded.

In addition to the hair fragments, purple fibers were also found within this impact area of the windshield. These fibers were identified as cotton. Since the laboratory, at this point, had only received the blue jeans and shoes from both individuals, the officer was contacted for a complete list of all evidence collected, with a specific request for items which were purple. Other impact areas on the windshield were also examined, including one area near the rearview mirror, which was also created by a force originating from the interior of the vehicle. No further trace evidence, however, was located at this site. An additional impact area was examined on the passenger side of the windshield. It was determined that this impact was made by contact from the outside of the vehicle, however, not the inside. Further examination of this area would yield no evidence, which could assist in placing the occupants in their respective seats.

The officer had also submitted both brake and accelerator pedals in the hopes of finding "shoe tread patterns". However, these patterns are rarely found. It would be more likely that trace evidence transferred onto the pedals.

Both pedals were examined, and no tread patterns were found, although, loose fibers were collected from the accelerator pedal. No foreign fibers were collected from the brake pedal. Nothing was located, however, which could be associated with severe contact or impact. Nevertheless, the shoes of both individuals were also examined. Even though the shoes had not deposited transfers onto the pedals, the pedals or other surfaces within the vehicle, may have transferred materials to the shoes themselves. Many circumstances can develop in which the examination of shoe soles would be important. Most shoes have a soft sole, which lends itself well to preserving transferred evidence. It is therefore extremely common to find various types of trace evidence in tennis shoes.

In this particular case, the shoes from both individuals had been submitted. Tommy had been wearing a pair of boots, of which the soles were leather, making the transfer of trace evidence less likely since this is a firmer substance than most rubber soles. However, they were nevertheless examined for any foreign materials. Nothing was found in the soles of the boots; however, a smear was located on the left toe area. This was collected, since the transfer might also give a clue to the path which Tommy had taken inside the vehicle or in exiting the vehicle. The polymeric smear was analyzed using FTIR spectroscopy and a "fingerprint" pattern was identified for this foreign substance. For comparisons, further standards from the vehicle were still needed however. Once again, the officer was contacted for a complete list of all evidence collected, with a specific request for items which Tommy may have come in contact with, in order to deposit the transferred polymeric material.

Jim's shoes provided much more information. He had been wearing a pair of dock-style shoes with a soft, rubber sole. The soles of both shoes showed several areas of contact: the right shoe had a large area in its outer-heel corner, which was scuffed, causing numerous fibers to be caught in the soft, rubber sole (see Figure 8.7). The left shoe also had an area on its inside-heel where fibers had been caught in the soft rubber sole (see Figure 8.8). These fibers had to have been deposited during severe impact in order to mesh into the soft polymer of the shoe soles. The foreign fibers from both shoes were collected. Once again, the officer was contacted for a complete list of all evidence collected, with specific request for any carpeting or floor mat samples.

From the officers' initial submission, valuable information had been gained to determine the possible driver of the Mercedes. The left lower speaker grill had impacted blue cotton and white cotton fibers which could have come from either Jim's or Tommy's blue jeans; only Jim's blue jeans, however, had a corresponding transfer of the blue thermoplastic material onto the blue jeans. There was much more information that could still be determined from remaining items, as many questions were still unanswered. Purple fibers collected from the windshield did not have a possible source. Various fibers collected from

Figure 8.7
Shoe sole with foreign fiber smears.

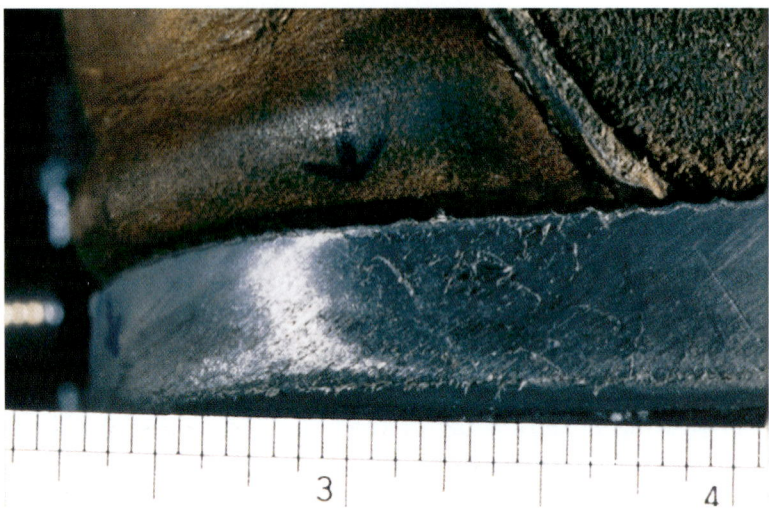

Figure 8.8
Photograph, shoe sole with foreign fibers.

Jim's shoes did not have a source. Finally, foreign smears on Tommy's boots did not have a source. From the submission of additional items, which had been secured, but left in the police department's evidence room, new information could still be obtained. Thorough scene work and excellent communication between the analyst and officer allowed this investigation to continue.

A second submission of evidence was brought to the Trace Evidence Unit. Shirts from both occupants were submitted for comparisons. Could this be the source of the purple fibers? In addition, the driver's side floor mat and carpeting was submitted. Carpeting which was also exposed in the center of the vehicle was sampled and submitted. Could the fibers and polymer material found on the shoes and boots have originated from these carpeting samples? Further analysis could begin to obtain the answers to questions still unresolved.

Is it possible that the purple fibers removed from the windshield could have come from the driver's shirt? Both Jim and Tommy had on dark colored, long sleeve shirts. The color of Jim's shirt, however, proved interesting. It was purple. Tommy's shirt was once again, a blue jean material. Closer examination of Jim's purple shirt revealed several areas where the fibers had been crushed, which is indicative of an impact area. One of these areas was on the right sleeve, near the cuff. A known sample was collected from the shirt for comparison with the fibers that had been removed from the windshield. The known sample from the shirt was purple cotton. The various shades of coloring of the purple dye in the purple cotton were similar to that found in the fibers from the windshield. A closer examination of Tommy's shirt also revealed additional information: on the right sleeve, an area of crushed fibers was found. While no foreign material was found in these crushed fibers, there may have been areas in the vehicle containing fibers, which were transferred from this contact. The list of collected evidence was once again reviewed. The officer was contacted for additional evidence collected, particularly for any parts of the vehicle which might contain damaged areas that Tommy may have come in contact with during the accident.

Examination of the carpeting samples proved just as interesting. Photographs of the vehicle revealed the Mercedes had several layers of carpeting. To accurately compare the fibers which had been collected from Jim's shoes, samples were needed from all carpets. First, a navy carpet was present over the entire flooring area. A sample of this carpet was collected from the driver's side of the vehicle, and was submitted to the laboratory. In each seat area, solid navy floor mats had been added. This floor mat from the driver's side was also submitted for comparisons. On the driver's side floor, an additional mat had been added as well. This navy designer mat contained a lighter navy panel in the center with gray lettering: "Mercedes Benz". This floor mat was also submitted for comparisons (see Figure 8.9).

Samples were taken from the carpet, the panel, and the lettering for further testing.

Various microscopic characteristics of fibers may be identified by examinations using a polarized light microscope and a fluorescence microscope. The overall color, shape, and diameter of the fiber are noted. Any variation in color, and whether the color is dyed or surface dyed, should be determined. Other

Figure 8.9
Driver's floor mat.

important identifying characteristics that may be observed using a polarized light microscope are a fiber's optical properties, such as its refractive index, birefringence, sign of elongation, and pleochroism. The refractive index of fibers should be determined for both the parallel and perpendicular axes, as these values are different, and also provide information about the birefringence and sign of elongation of the fiber. Various dyed fibers will also exhibit pleochroism, which is the differential absorption of light by an object when viewed at different orientations relative to the vibration direction of the polarized light (i.e. the fiber may exhibit different colors in different orientations). Inclusions in the fiber, such as optical brighteners and delustering agents, and the size and shape of these particles, should also be noted. The cross-section of a fiber is also an important characteristic and can be determined using the polarized light microscope. The cross-section may be used to help establish a general fiber type and usage, and may come in a wide variety of shapes: round, bilobal, trilobal, multilobal, bell, dogbone, triangular, hollow, and irregular.

Some fibers will also fluoresce as a result of the fiber itself or from dyes and additives in the fiber. This can be quickly determined using a fluorescence microscope. Examinations should be performed using a variety of excitation energies and barrier filters. The color and intensity of the fluorescence, or the absence of fluorescence, should be noted. Instrumental analysis of fibers can provide additional information for man-made fibers. FTIR spectroscopy is valuable in determining the generic class of a man-made fiber, and may be useful for separating fibers within a generic class. Information on fiber additives or dyes can also be distinguished at times using FTIR spectroscopy.

Polarized light microscopy, fluorescence microscopy, and FTIR spectroscopy were the only instruments available at the time for analysis on these samples. Analytical testing revealed the fibers from the carpeting, floor mat, and designer floor mat were all man-made fibers. The carpeting sample from the flooring and the plain floor mats contained fibers, which were similar. The designer floor mat, composed of the background, the panel, and lettering, was different from the other fibers. Even though the three fiber samples from the designer floor mat were also nylon, the microscopic characteristics were different. Since the designer floor mat was on the top, exposed to the driver, and the floor mat on the passenger side were microscopically different, it now became important to test the fibers removed from Jim's shoes to determine if they were identical to any of the known fiber samples. Closer examination of these fibers removed from the shoes revealed two different shades or colors of fibers. Could it be that some of the three different fibers from the designer floor mat had been transferred? After complete analytical testing was performed, the fibers removed from Jim's shoes showed no differences from the fibers taken from the two of the sections of the designer floor mat. The fibers removed from the shoe soles were identical to the fibers from the panel and the lettering areas of the designer floor mat. Further examination of the designer floor mat revealed several polymeric smears. The smears were removed and compared to the soles of Jim's shoes. Using FTIR spectroscopy, it was determined that the smears and the soles of Jim's shoes were composed of the same color and general polymer type. Once again, as with the speaker grill and blue jeans, material had been transferred to and from both possible sources.

Communications between the analyst and the officer continued to allow the investigation to continue. A third submission of items was delivered to the Trace Evidence Unit. The passenger interior door panel had visible areas of contact. Could these contact areas have been created by Tommy's shirt rubbing up against them, causing the crushed fibers? Additionally, the airbag and steering wheel were submitted for analysis. Since the airbag had deployed, contact with the airbag and perhaps the steering wheel, would have been made by the driver. The possibility of trace evidence transfer to these items needed to be determined.

The passenger interior door panel not only had visible areas of contact, but upon closer examination, an actual fabric impression could be seen. A fabric impression, or weave pattern, can be left on an item after contact with a fabric. A twill-weave pattern was visible in the area, which had been scraped. In this weave, considered the most durable type of weave, the fill yarns are interlaced with the warp yarns in such a way as to form diagonal ridges across the fabric. Common fabrics using the twill weave are: jean, denim, dungaree, gabardine, whipcord, houndstooth, and melton. The fabric of both pairs of blue jeans was woven in a twill-weave pattern, as was the denim shirt. Foreign fibers were also contained in the fabric impression. These fibers were removed, and found to be blue cotton and white cotton. It was now possible that the foreign fibers on the interior passenger door panel were, therefore, similar to three sources: Jim's blue jeans, Tommy's blue jeans, and Tommy's shirt.

Could the airbag and steering wheel also have foreign fibers from the driver's clothing? The airbag was examined using a stereomicroscope. Several foreign fibers were collected. Blue cotton, white cotton, and purple cotton were identified. These foreign fibers were compared to samples previously examined. The various shades of coloring of the blue dye in the blue cotton were similar to that found in all of the blue jean samples: Jim's blue jeans, Tommy's blue jeans, and Tommy's shirt. The various shades of coloring of the purple dye in the purple cotton were similar to that found in the fibers of Jim's shirt. The steering wheel was also examined using a stereomicroscope. Additional blue cotton, white cotton and purple cotton fibers were also found. These fibers were compared to the previously submitted samples, with similar results. Had enough of the questions been answered to determine a possible driver of the Mercedes the night of the fatal accident?

COURT PROCEEDINGS

After presenting the officer's accident reconstruction findings and the laboratory results before a grand jury, an indictment was received. Jim was formally charged with the murder of Tommy on the night of the accident. The prosecution had presented a sound case during the grand jury proceedings. Jim's family was prominent in the community, and they were able to secure an ex-prosecutor as the defense attorney. Experts were also called in to re-examine the accident reconstruction work, as well as the laboratory findings. Both of these experts requested access to the vehicle, all items collected, all items examined, and any reports from the officer or analyst. Court orders to this effect were signed almost twelve months after the accident.

The vehicle had been stored, uncovered, in the impound lot at the police department. All items collected, which the laboratory did not analyze and those analyzed, were stored in the police department evidence room. The microscope slides, which had been made by the analyst during analytical testing, remained at the laboratory. Both defense experts were allowed access to all items. However, the results of the defense experts were quite different from those of the officer and the laboratory analyst.

In essence, the defense expert, upon re-examining the accident reconstruction findings reached conclusions similar to the officer's; however, he positioned the individuals in reverse: Jim was placed in the passenger's seat, with Tommy in the driver's seat. No detailed assessment as to why the individuals were placed in that manner was included in his report. His evaluation of the vehicle and occupant motions was similar to the officer's. Only slight differences were noted as to the speed of the vehicle.

The re-examination of the trace evidence was a different matter. The defense expert had requested access to the vehicle, all items collected, and all items examined, in addition to the report from the analyst. Upon evaluating the findings of the defense expert, however, it appeared that only a portion of these items were taken into consideration during his assessment. Even though the Mercedes had been stored uncovered at the impound lot during the twelve months following the accident, the defense expert collected and examined items from the vehicle; including hair that was collected from the passenger seat belt harness. In addition, evidence was collected from items, which had been previously examined. Hair was collected from the passenger's side of the windshield, and the driver's side floor mat. Fibers were collected from the brake pedal and Tommy's boots. An imprint had also been found on Tommy's T-shirt. Also contained in the report was the defense expert's assessment of the bodily injuries to both individuals. His final report stated the conclusion that Tommy was the driver; Jim, the passenger. Was it possible that the officer and analyst had missed all this evidence? Could they had reached the wrong conclusion?

Since equal discovery had been granted, the prosecution received copies of reports from both defense experts; accident reconstructionist and trace analyst. The defense accident reconstruction report varied slightly from the officer's conclusions, the main difference being the speed of the vehicle. The occupant and vehicle dynamics were similar. The defense expert's placement of the individuals seemed to be determined by the findings of the defense trace expert and the individuals' injuries. The officer's placement of the individuals was made from the assessment of the laboratory reports, in addition to the injuries sustained by each individual. While determining such similar results for the occupant and vehicle dynamics, how could such different conclusions be reached for the occupant placement? The only difference was the trace evidence.

Upon closer review of the defense trace expert's report, no mention of any evidence, which had been collected and analyzed by the laboratory analyst, had been taken into consideration in his results. Even though he had access to these items, he chose to weigh his findings only on what he himself had collected. Only "new" evidence had been used to base his opinion. Could the analyst have overlooked all these items?

Review of the analyst's notes show that the seat belt harness had never been submitted for examination. Questions were raised as to whether hair collected twelve months later from the uncovered vehicle should be considered. The analyst agreed to evaluate and compare this hair if requested, since as an expert witness, opinions as to its appropriateness may be given during testimony. The defense expert had also found hair on the passenger side of the windshield. The entire windshield had been examined by the analyst and her notes indicated no hairs were found on the passenger side, only on the driver's side. The defense expert had also collected additional hair from the driver's side floor mat. Since the vehicle had flipped, only hair, which was imbedded, as that in the windshield, had been collected by the analyst for examination. Hair, which was found on the floor mats was loose, and could have been deposited at any time. The analyst agreed to evaluate and compare this hair if requested, since during testimony, further explanation could be given so that the jury could understand the value of this evidence.

In addition, fibers had been found by the defense on the brake pedal; the analyst's notes indicated only a small fragment from a feather was seen on the brake pedal. The defense expert found fibers on Tommy's boots; the analyst's notes indicated no fibers were found on the boots. Fibers had been removed and analyzed from Jim's shoes however. The defense expert did not collect any of these, even though it was noted by the analyst that additional fibers remained on the shoes. An imprint had also been identified by the defense expert on Tommy's T-shirt. Although the T-shirt had not been submitted for examination, imprints from vehicle parts should only have been considered on the outer-most layer of clothing. Nevertheless, the analyst agreed to evaluate and compare this imprint if requested. As with the other evidentiary assessments, further explanation could be given during testimony about imprints so that the jury could understand the value of this evidence.

The analyst had completely and accurately analyzed all of the items. The analyst had covered all bases. The defense was taking a different road. They chose to ignore the evidence, which had been collected, analyzed, and presented in the laboratory results and develop their own theory. Would this work before a jury? To know the answer to this question, a better understanding of the "new" evidence was needed. The prosecution decided to re-evaluate the evidence presented by the defense.

The hair, which had been collected from the seat belt harness, reported by the defense as being similar to Jim's known standard, contained two head hair fragments. These hair fragments were packaged on a microscope slide, under a cover slip which was held by tape. This is not the preferred method of mounting a hair sample. The fragments were re-mounted using a mounting medium close to the refractive index of hair, which allows the analyst to view the microscopic characteristics of the hair. The examination revealed characteristics in one hair fragment similar to a known standard; it was, however, the known standard provided from Tommy. The second hair fragment was different from both known standards. As is common practice in the laboratory, a second opinion was provided by two other hair analysts. They agreed with the assessment made by the analyst, not the defense expert.

The hair fragments collected from the windshield and the driver's floor mat were packaged in a similar manner. After re-mounting the samples, they were found to be too limited for comparative purposes. Sections of the hairs had been mounted on Scanning Electron Microscope (SEM) stubs. These were not analyzed, since the laboratory does not perform SEM testing on hair samples, unless heavy metal poisoning is suspected. All fibers, which had been collected by the defense expert, had also been mounted on the SEM stubs. The packaging of the stubs had allowed for possible contamination, however, and the sticky areas on the stubs now contained a great amount of fibers. Since contamination had possibly occurred, without any assurance that the stubs contained only those fibers collected and analyzed by the defense expert, further analysis was pointless. Tommy's T-shirt was also examined for reported imprint areas. Several areas of damage were found, however, the only area of possible transfer contained both blue cotton and white cotton fibers. This would be consistent with the T-shirt striking the outer layer of blue jean shirt. No transfer had been made with a vehicle part, as portrayed by the defense. It was now time to see what the jury would believe.

The Assistant Commonwealth Attorney methodically laid out all the testimony, presenting witnesses who testified as to their knowledge of the facts. The officer presented his reconstruction of the accident, using the vehicle dynamics and occupant kinematics, to testify, in his expert opinion, as to who the driver of the Mercedes was on the night of the fatal accident. The trace analyst also presented her evidence. On the left lower speaker grill, impacted foreign blue cotton and white cotton fibers were found. Blue cotton and white cotton fibers were also found on the air bag and the steering wheel. The blue jeans from both individuals and also the blue jean shirt from Tommy were composed of blue cotton and white cotton, which was similar to these foreign fibers. On Jim's blue jeans, an area of impact was found on the left knee. This contained a blue polymeric material which was similar in color and general polymer type as that of the speaker grill. Foreign purple cotton fibers were found on the driver's side of the windshield. Purple cotton fibers were also found on the air bag and steering wheel. Jim's shirt was composed of similar purple cotton fibers. Foreign impacted fibers were found in the soles of Jim's shoes. These fibers were identical to the two different fibers composing the designer floor mat on the driver's side of the vehicle. The designer floor mat contained foreign polymeric material which was similar in color and general polymer type as that of Jim's shoe soles. The passenger's side interior door panel contained an area of twill-weave pattern impact which contained foreign blue cotton and white cotton fibers. This is consistent with both pair of blue jeans and also the blue jean shirt from Tommy. These results reinforced the officer's opinion, delivering the jury one possible conclusion: Jim was the driver.

The defense then presented their case. The accident reconstructionist testified to findings similar to those of the officer. The defense analyst, however, presented a different picture. SEM photographs were presented, showing the similarities of the hair fragments collected from the passenger side windshield and the seat belt harness to Jim's known standard. SEM photographs were displayed showing the similarities of the hair fragments from the driver's side floor mat to Tommy's known standard. Who was the jury to believe?

During the rebuttal by the prosecution, several questions, which arose during, direct or cross-examination testimony were cleared up. Then, the Assistant Commonwealth Attorney asked about SEM: why hadn't the laboratory analyst examined the hair using this technology? The analyst explained the current methods of hair analysis accepted within the scientific community, and the values of each. The attorney then asked the analyst to view the SEM photographs presented by the defense expert. Even though her laboratory didn't normally perform hair analysis using SEM, and her microscopic opinion differed, he asked her to reach a conclusion solely from the defense expert's SEM analysis and photographs. Everything was on the line. If she agreed with the defense expert's assessment of the photographs, Tommy was the driver. If she disagreed, it would become a "who do you believe" jury problem.

While evaluating the photographs, she found that the images of the scale patterns of the hair and elemental data presented were very similar. Scale patterns of human hair, however, would almost always be similar. Likewise, the elemental composition of hair is generally the same. If these photographs, as the prosecuting attorney asked, were more or less deciding the case, there had to be something else. And there was. Upon further evaluation, the analyst noticed that the magnification of the hairs varied from one photograph to the next. The defense expert was, in essence, asking the jury to evaluate and compare hairs at differing magnifications. He wanted them to formulate opinions from his photographs even though they were created under different circumstances.

After giving the analyst time to evaluate the photographs, the prosecutor repeated his question. She proceeded first by quickly discussing SEM analysis. The instrument and the general methods used for elemental analysis were explained. Then SEM analysis of hair was discussed, while stressing the importance of sampling and instrument operating parameters. Next, each photograph was described. The information gained from each image and elemental analysis was explained. It now was time to describe the comparisons of the photographs and the results. The answer was simple: this could not be done! Comparison of hairs, which had been examined at one magnification – whether it is using a microscope or a SEM – should not be compared to hairs examined at a different magnification. The analyst explained this to the jury and why conclusions should not be reached under these circumstances. These photographs should not be compared, resulting in no conclusion being made. The defense expert had asked them to compare apples to oranges.

After 4 days of testimony, the closing arguments were made and the jury was sent to deliberate. They quickly reached a verdict. They believed the evidence proved, beyond a reasonable doubt, that Jim was the driver. The defense appealed the guilty verdict, but the Kentucky Supreme Court upheld the conviction, and after almost 4 years from the date of the accident, Jim was sentenced to 7 years in jail for Tommy's death. An additional 18 days and a $200.00 fine was imposed for driving under the influence.

Even after the conviction had been upheld, Jim still maintained his innocence. His family supported his belief and eventually convinced a law school in New York to investigate the case. They, in turn, worked with an agency established to protect and preserve human rights. Initially, all notes, sketches, interviews, and reports, concerning the case were evaluated. All records relating to the evidence and the security of the evidence, including the vehicle were requested. Transcripts of all court proceedings were evaluated. For the expert witnesses, the officer and the analyst, additional information about their qualifications, degrees and training, were requested. An overview of all operations, procedures and instrumentation utilized by the laboratory was also requested. Numerous contacts were made by the agency with the officer and analyst. Many additional contacts were made by the agency, questioning their qualifications and training. After many months, the investigation concluded quietly, with no further action being taken.

ACKNOWLEDGEMENTS

The author would like to thank Kevin Robinson for his never ending pursuit of answers. Without his dedication to his career and justice, much of the evidence in this case would never have been analyzed. Also, thanks to his agency, Lexington-Fayette Urban County Police Department, for allowing him to share information from the casefile so that the complete story might be written. An last but not least, special thanks to Lara Mosenthin for her help in proof reading.

CHAPTER 9

MY ROOMMATE IS USING THE REFRIGERATOR

Max M. Houck

Director, Forensic Science Initiative, West Virginia University, Morgantown, WV
Executive Director, Institute for Cold Case Evaluations, Corp. (ICCE)
Morgantown, WV

INTRODUCTION

The body of a Saudi Arabian doctor was found in a refrigerator near a garbage dumpster behind a sporting goods store; she had been strangled. A number of evidence types, including rope, carpet fibers, and duct tape, were used in the investigation and prosecution of her murder. The murderer was her roommate, who had found a new tenant to share his apartment, and he decided to remove the doctor permanently when she protested her "eviction."

None of the evidence found on the refrigerator was, by itself, damning; after all, it was the murderer's appliance and had been seen in his apartment. But, given his statements, the statements of others, and the context in which the crime occurred, the physical evidence led the jury to their final conclusion. As is the case with many criminals, the murderer boxed himself into an inescapable corner through his own actions.

THE CRIME SCENE

Three construction workers arrived for work near a shopping mall and noticed a light-colored refrigerator (Figure 9.1) taped shut with white tape and tied with black and orange rope behind one of the stores. They thought this was odd and, with their curiosity peaked, tipped the refrigerator so that the door faced upwards, cut the rope and tape, and opened it. To their horror, they discovered the body of a middle-aged woman, bound, clothed, with a box on over her head (Figure 9.2). A number of other items were in the refrigerator, such as cotton towels, pieces of cardboard with orange "Air Canada" tape on them, and paper bags.

At autopsy, lividity was noted in two patterns, on her right side and back, consistent with her positioning. Blood in her nasal passages, periorbital and oral petechiae and bruising led the medical examiner to determine the victim's cause

234 TRACE EVIDENCE ANALYSIS

Figure 9.1

The refrigerator where the victim was found.

Figure 9.2

The area near the shopping mall where the refrigerator was found.

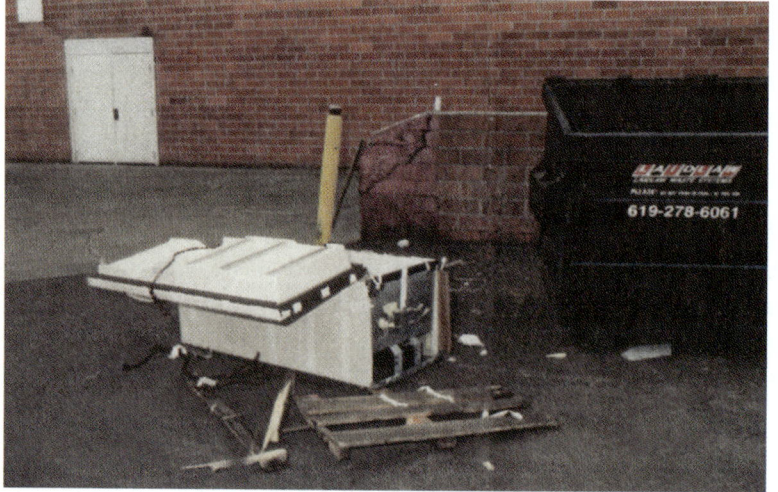

[1] All names have been changed.

of death to be "strangulation by another person(s)." When her fingerprints were checked against the state's fingerprint file, the victim was identified as Sarah Mahtah,[1] a licensed physician. Police had suspected Mahtah's identity as one of the boxes removed from the refrigerator had a mailing label bearing her name, an address, and telephone number. That information led police to a law office and Mahtah's employer, who provided police with Mahtah's home address. It turned out to be a high-rise apartment building where Mahtah sublet a penthouse apartment with the lessee, Tony Butters.

When Butters was contacted by police, he admitted that he knew Mahtah, but told them that she had recently moved out. Butters told police that he and Mahtah were "incompatible" and that she had agreed to vacate on the 19th of the month; a new roommate, also a female, had moved in with Butters on the 20th of the same month. Butters said he had watched football games on television the entire day while Mahtah and some man he had never seen before moved her things out. Police informed Butters as to the nature of their visit and began, with Butters' consent, to search the apartment. Butters' offered saliva samples and his fingerprints; various other known materials were also collected, such as carpet fibers from where the refrigerator sat. Additionally, the impression of the refrigerator was still visible in the carpeting; the investigators took measurements.

While the police conducted their search, a woman entered the apartment and identified herself as Jocelyn Manning, Butters' current roommate. Detectives pulled Manning aside and spoke with her about the circumstances surrounding the time when she had first viewed and then subsequently moved into the apartment.

When Manning first viewed the apartment to determine if she wanted to live there, she had noticed that the apartment was clean and oddly bare of furniture, with the exception of a second refrigerator in the living room. She saw clothing and boxes in the room where she would live; Manning also saw women's cosmetics and clothing in the bedroom's bathroom. Manning asked Butters about the refrigerator and he responded that it was his roommates' and she was the only one who used it. Butters assured Manning that the refrigerator would be gone by the time she moved in. When she moved in the next day at about 7 p.m., Manning noted that the refrigerator was gone but the carpeting where it had sat was moist; Butters said he had shampooed the carpet. She thought this odd because Butters had told her that the carpeting was only 9 months old and it had looked quite clean to her. Detectives asked Manning to describe the refrigerator and she said it was about 5 ft. high, 24 in. wide, and "almond or light yellow" in color.

When she moved in, Butters said he would help her move her things and he produced a furniture dolly from his apartment. She asked him if he had gotten it from the rental office and he had said, no, he just had it. The dolly was a rectangular wooden frame, with casters on the bottom and carpeting on top to cushion whatever was being moved. A later interview with the rental office staff revealed that the dollies were normally checked out on a timesheet but, often, tenants would just borrow them and not sign them out.

Detectives told Manning why they were there and the nature of the investigation. Manning decided to move out immediately. Police escorted her to the parking garage, where they noticed something familiar: two parking spaces had been marked out with white tape that looked similar to the tape that had bound Mahtah's hands and sealed the refrigerator. Manning said that the parking spaces were Butters'. As the investigators looked over the area, they also found a piece of cardboard with orange "Air Canada" tape on it.

As the investigation continued, it became clear that Mahtah had disappeared under curious circumstances. Her car was located but had no clothing, toiletries, or boxes in it. Her brother told police that Mahtah communicated with him frequently and on the 16th of the month of her disappearance she told him that she had an argument with her "landlord." Mahtah said that she would be leaving at the end of the month. He then spoke with her again on the 18th of the month and during that conversation she had not indicated that she had changed her plans to move any earlier than the end of the month, nearly 2 weeks away. A search warrant was obtained and Butters' car, apartment, apartment building and storage area were searched.

A number of results from the local forensic science laboratory indicated Butters' involvement in Mahtah's death. The white tape from the refrigerator and the victim's wrists was found to be similar in manufacturing characteristics to the white tape from a wooden pallet in Butters' storage area and to the white tape on the floor of the garage delineating Butters' parking space. The orange "Air Canada" tape from the refrigerator and from the storage area also appeared similar. Black and orange rope from the refrigerator exhibited similar manufacturing characteristics to black and orange rope from moving carts in the apartment manager's office in the building where Butters and Mahtah lived. Light-colored carpet-type fibers found on the tape from the refrigerator appeared similar microscopically to carpet fibers from Butters' apartment.

The impression in the carpeting was measured and compared with the bottom of the refrigerator; the refrigerator was determined to be a potential source for the impression in the carpeting. A fingerprint matching Butters' was found on the refrigerator.

While this evidence was enough to arrest Butters, some additional work was needed. The local forensic science laboratory did not have the capability to analyze the fibers from the carpeting and the ropes comprehensively. They had employed the analytical tools they had in their laboratory, microscopy and infrared spectroscopy, but did not have the ability to analyze the fibers for dye or color information. Certain items of evidence, specifically the collected fibers and rope from the refrigerator and known carpet fibers from Butters' apartment and rope from the furniture dollies, were sent to the FBI Laboratory in Washington, DC for additional analyses. The letter requested fluorescence microscopy and microspectrophotometry, as well as "any additional information you may be able to provide." Although local laboratory intended to increase the specificity of the fiber and rope associations through color analysis, they got more than they asked for.

THE ANALYSIS

THE CARPET FIBERS AND COLOR

The vast majority of fibers used in commercial applications are given color. Color is one of the most important properties in the comparison of fibers and is a critical test in any analytical scheme for fibers. Synthetic dyes and pigments belong to 29 different chemical categories with more than a dozen different application methods. Even seemingly simple dyes might require between eight and ten processes to convert the raw materials into a finished dye suitable for use in coloring textiles. Given that the total annual production of any particular dye might not amount to more than 10 tons and that small process batches are becoming the rule in the dyeing industry (Apsell, 1981), color becomes a powerful discriminating property. The selection of dyes is based upon many factors that, while not based on the final desired color, nevertheless affect the textile's appearance (Park and Shore, 1999). Color becomes particularly significant when the gamut of colors is considered: literally, millions of shades are possible in textiles. When these colors are spread out across the range of garments and carpeting produced in any one year, and "multiplied" by the number of garments and carpets produced in previous years, the importance of color cannot be underestimated. A graphic view of the manufacturing process quickly shows the dynamic nature of this system and many of these factors can affect color (Figure 9.3).

Individual fibers can be colored before being spun into yarns; once spun, the yarns themselves can be dyed. The finished fabric can be dyed before or after its construction. Color can also be applied to the surface of a fabric by printing or resist dyeing methods. Fading and discoloration due to environmental effects may also add increased significance to a fiber association. In short, variation is the forensic fiber examiner's friend and provides the basis for their analysis and comparison of textile evidence.

There are a variety of methods for characterizing either the color of and/or the dye(s) in the fibers. They fall into three major categories: visual, chemical, and instrumental. The visual method is the simple observation and comparison of the fiber colors by use of the aided eye. Visual comparison is easy, fast, and non-destructive. It is a crucial first step in any fiber comparison, as many, otherwise similar fibers can be excluded from consideration simply by looking at their color.

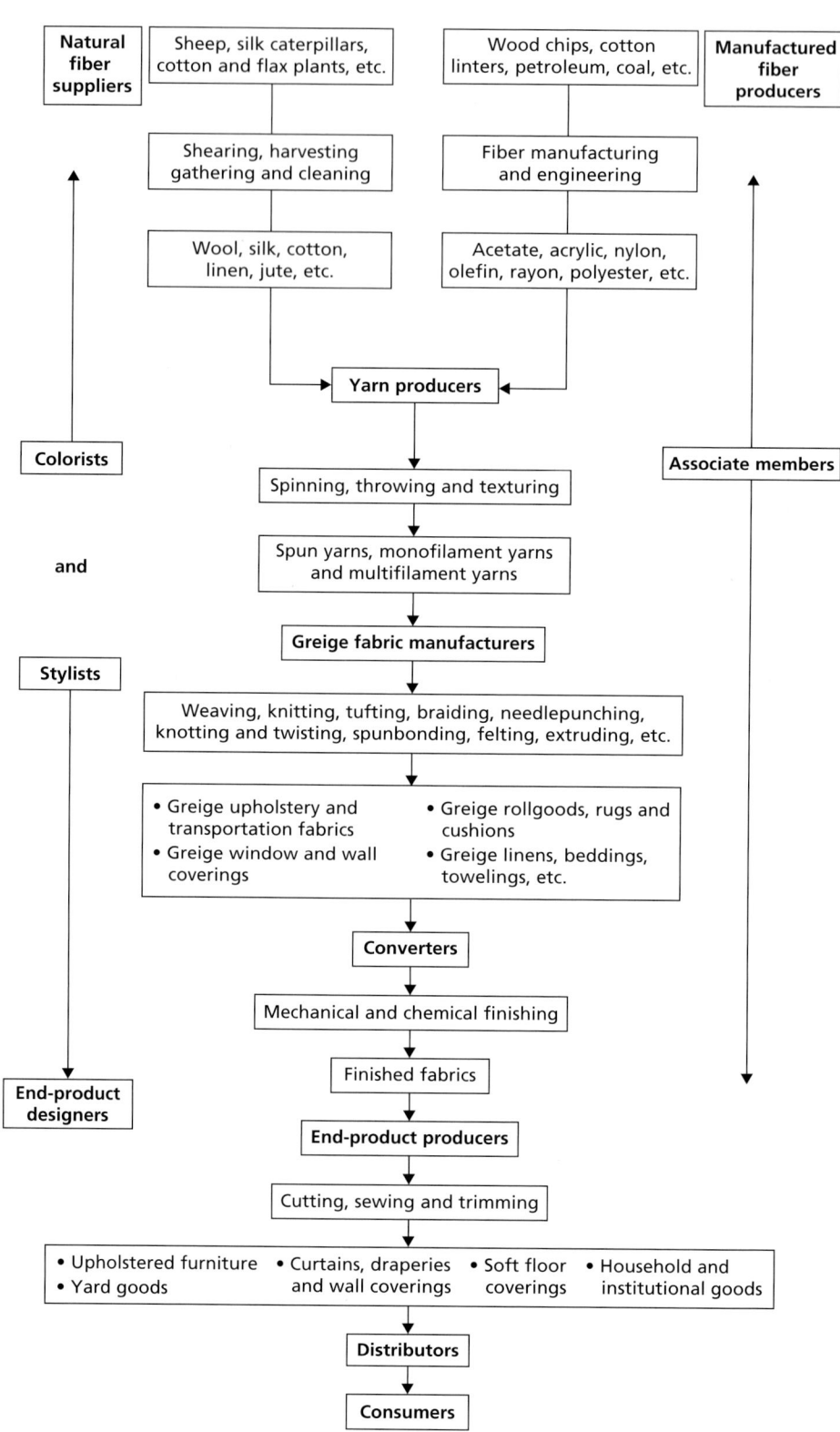

Figure 9.3
The manufacturing process for textiles. Note that the coloring of textiles can occur at nearly any point in the process and this will affect the visual nature of the end product, both grossly and microscopically (From Yeager and Teter-Justice, 2001 with permission).

Visual examination can also involve various types of analytical microscopy and, for color, this can take the form of fluorescence microscopy. Fluorescence is a type of luminescence, which is the property of certain substances that, when irradiated with UV, violet, blue, or green light, emit radiation of their own whose wavelength is longer than that of the exciting light (Stoke's Law). Luminescence can be subdivided into two phenomena. The first is phosphorescence, where the secondary emission persists for a time after the excitation has ceased. An example of phosphorescence is the paint on clock and instrument dials, which contains a tritium isotope and phosphor in the resin-solvent paint base. These paints have a half-life of 12.5 years and are perfectly safe because they emit only low secondary beta radiation and the plastic or glass cover effectively blocks the emission (Brady and Clauser, 1991).

Fluorescence, the second type of luminescence, exhibits emission only as long as the excitation continues: when the excitation stops, so does the emission. Fluorescence can be further categorized as either primary or secondary fluorescence. Unstained materials that emit fluorescent light when they are excited by short wavelength light exhibit primary fluorescence. Many specimens, however, do not exhibit any primary fluorescence at all. If they have color imparted to them, they may then fluoresce.

This coloration is usually intentional, such as staining tissues for histology or coloring textiles for consumer use, although some accidental staining can certainly produce fluorescence; semen, for example, fluoresces at about 450 nm. The dyes that cause fluorescence are called fluorochromes; those used for biology or medical research are known to fluoresce and are used specifically for that purpose. Textile dyes that fluoresce are normally not used intentionally for that reason and it is a functional accident that forensic scientists exploit. In certain studies, textiles have been labeled or tagged with fluorescent dyes, such as fluorescein isothiocyanate, for ease of detection (Houck *et al.*, 1999).

The wavelength of the emitted fluorescence light is longer than that of the exciting radiation. In other words, radiation of relatively high energy falls on a substance. The substance absorbs and/or converts (into heat, for example) a certain, small part of the energy. Most of the energy which is not absorbed by the substance is "emitted" again as fluorescence. The fluorescence radiation has lost energy and its wavelength, therefore, will be longer than that of the exciting radiation. There is an inverse ratio for radiation energy to wavelength: shortwave gamma rays or X-rays, for instance, penetrate deeper than longwave light and heat rays. Consequently, a fluorescing substance can be excited by near-UV radiation and its fluorescent components (called fluorophores) emit in the visible range (Figure 9.4).

The basic components needed for fluorescence microscopy are:

- a light source which supplies enough radiation of a wavelength required for fluorescence excitation,

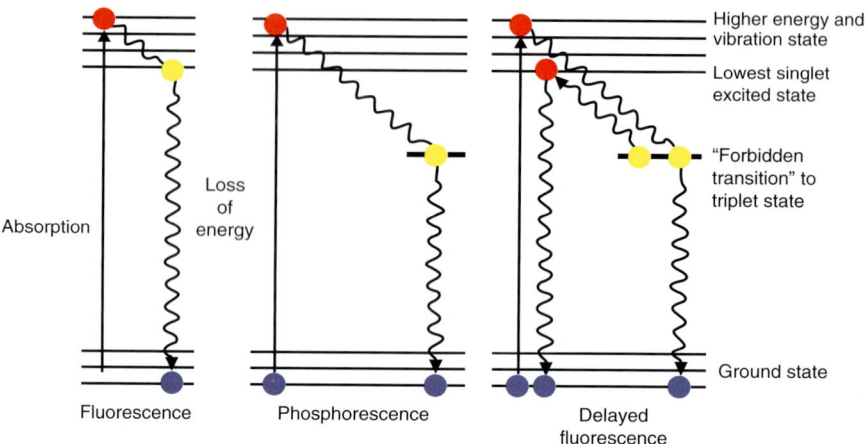

Figure 9.4
Fluorescence diagram
(Courtesy Olympus America, Inc.).

- an excitation filter which transmits only the exciting radiation,
- a fluorescing specimen, and
- a barrier filter which suppresses excessive exciting light which is not absorbed by the specimen and would interfere with fluorescence.

Light enters the microscope system and an excitation filter cuts off or isolates certain regions of the spectrum. Within the visible spectral range a "colorless" filter transmits all spectral colors uniformly and with high output. Without losses due to reflection, its transmittance $D = 100\%$ for all wavelengths and the transmittance "curve" is a straight horizontal line. If certain regions of the spectrum are attenuated or completely absorbed the results are filter curves, plotting absorption versus wavelength.

The principles of optical microscopy apply as much to fluorescence microscopy as to any other technique of optical microscopy (Rost, 1992). It follows that a florescence microscope differs from a microscope used for conventional absorption microscopy mainly in that it has a special light source and a pair of complementary filters (Figure 9.5). The lamp should be a powerful light source, rich in short wavelengths, e.g. ultraviolet or blue: high-pressure mercury arc lamps are the most common. A primary or excitation filter is placed somewhere between the lamp and the specimen. The filter, in combination with the lamp, should provide light over a comparatively narrow band of wavelengths corresponding to the absorption maximum of the fluorescent substance (fluorophore). To enable the comparatively weak fluorescence to be seen, despite the strong illumination, the light used for excitation is filtered out by a secondary (barrier) filter placed between the specimen and the eye. The secondary, barrier or suppression filter prevents the excitation light from reaching the observer's eye. This filter, in principle, should be fully opaque at

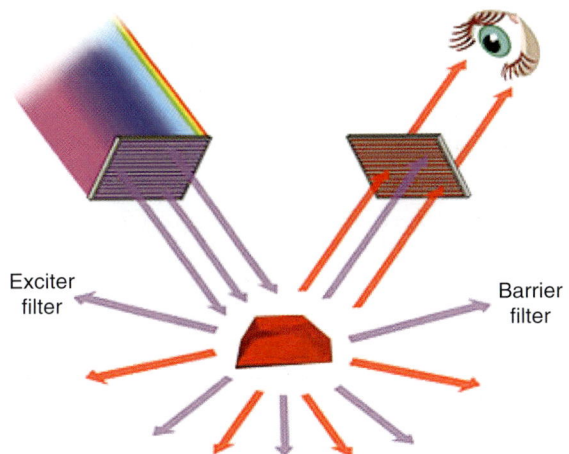

Figure 9.5
Schematic of fluorescence microscope.

the wavelength used for excitation, and fully transparent at longer wavelengths so as to transmit the fluorescence. The fluorescent object is therefore seen as a bright image against a dark background. The most important filter types for fluorescence microscopy are listed below:

- **Conventional glass filters**, which include longwave pass band filters and narrow the transmittance range.
- **Pass band filters** are high-performance filters with two sharp cutoffs within the spectral range, designated by the wavelengths of both cutoffs.
- **Shortwave pass band filters** are high-performance filters with sharp cutoff of the transmittance curve in the spectral range, cutoff designated by the wavelength of 50% transmittance in nanometers, thus wavelengths which are shorter than the cutoff pass through.
- **Longwave pass band filters** are like shortwave pass band filters but with a shortwave barrier range, thus wavelengths which are longer than the cutoff pass through.

A crucial component of this system is a dichroic mirror; these act essentially as beam splitters. Dichroic mirrors have multiple interference coatings which reflect over 90% of the short wavelength light while transmitting most of the longer wavelengths. On the return from the mirror, all the fluorescence is transmitted whereas the short wavelength exciting light is directed toward the light source and does not interfere with the image. The choice of dichroic mirror depends on the excitation maximum and the fluorescence spectrum of the fluorochrome. Excitation and barrier filters are fully effective only in certain combinations; the spectral combinations must not overlap. Microscopes are currently available that have a range of what are known as "filter cubes," which are specific

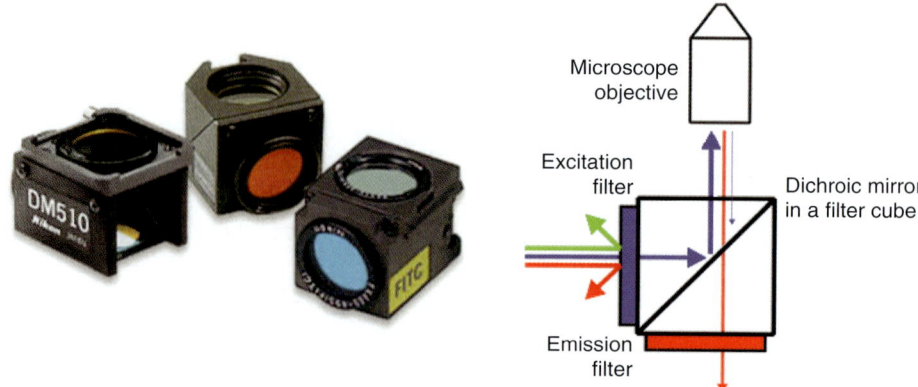

Figure 9.6
Fluorescence microscope with filter cubes and rotational housing (Courtesy of Olympus America, Inc. and Nikon, Inc.).

excitation–dichroic mirror–barrier filter combinations that are interchangeable within a special housing on the microscope (Figure 9.6).

Fiber/dye combinations may produce fluorescence of a particular intensity and color, both of which should be noted during the examination for each filter-mirror combination used. Fibers dyed with similar dyes should exhibit the same fluorescence characteristics, unless the fiber and/or dye(s) have been degraded by UV exposure, bleaching, or some other similar effect. Spectrophotometric methods have been attempted on fluorescent fibers with limited success (Hartshorne and Laing, 1971; Parts 1–3). The advent of solid-state detectors for spectrophotometry may change this, because the detectors require far less acquisition time than mechanical monochromaters and, therefore, less signal is lost.

Chemical methods, which include thin-layer chromatography and high-performance liquid chromatography, address the chemical nature of the dyes used to color the fiber. This latter statement is an important distinction: analyzing the color of a fiber is not the same as analyzing the dyes used to color that fiber. The examiner must make it clear which is being analyzed and not confuse the two, otherwise the interpretations offered could be markedly different.

Instrumental analyses are defined primarily by microspectrophotometry in the ultraviolet (UV) and/or visible ranges. The importance, significance, and methodology of this instrument have been published elsewhere.

The fibers were re-examined by light microscopy, polarized light microscopy, and then examined with fluorescence microscopy. The fibers were very light-colored, trilobal polyester carpet-type fibers with a slight distribution of small, oval-shaped delusterant particles. The settings on the fluorescence microscope were (all values are in nm): Green 2A cube (excitation 510–560, dichroic mirror 565, barrier 590), blue 2A cube (excitation 450–490, dichroic mirror 500, barrier 515), ultraviolet 2E/C cube (excitation 330–380, dichroic mirror 400, barrier 435–485), and a visible 2A cube (excitation 380–420, dichroic mirror 430, and barrier 450). The fibers exhibited slight yellow fluorescence under the blue 2A

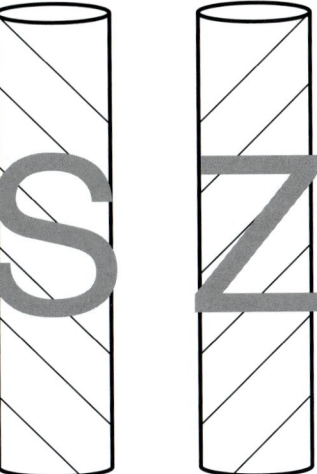

Figure 9.7
"S" and "Z" twist designations for ropes.

cube; no perceptible emission was seen under the other filter cubes. In this instance, no emission is just as important as seeing fluorescence. If fluorescence was seen in one sample but not in another, it would indicate that the two fiber samples had some differences in their fluorescence characteristics. This is essentially the same as, for example, one fiber sample emitting light blue under the ultraviolet cube and the other sample emitting violet under the same cube. If both samples, however, exhibit no fluorescence at all, then no conclusion can be drawn about their fluorescent characteristics.

The fibers from the scene and the known fibers from Butters' apartment proved to be too light for a reliable analysis by microspectrophotometry. Spectra that have an absorbance below 0.1 are generally considered to be unsuitable for a repeatable, meaningful analysis (SWGMAT Forensic Fiber Guidelines, 1999). For samples like this, it is particularly important to use fluorescence microscopy because it may be the only analytical signal available about the color or dye in the fiber.

THE ROPES AND THEIR CONSTRUCTION

Rope is constructed in two main ways, laid and braided; although there are many other variations on these basic patterns. The direction of twist is called the lay of the rope or just "the twist." Twisted cordage can be laid clockwise or counterclockwise. Twisted rope is described as S-laid or S-twist (clockwise) or Z-laid or Z-twist (counterclockwise) (Figure 9.7). The direction of the strands correspond to the slant of the middle section of either the capital letter "Z" or "S". The degree of twist is measured in crowns per inch, the number of "rises" created by the winding of the yarns.

The construction of stranded cordage is much the same whether it is made of natural or manufactured fibers. Individual fibers are twisted into yarns, the yarns are twisted into strands and the stands are twisted into cordage. The direction of twist generally runs opposite between construction elements: fibers are twisted in the same direction as the strands but the yarns are twisted in the opposite direction. By twisting cordage in this alternating fashion provides strength and keeps the line from kinking during use.

The other major construction type is braided cordage. Braided cordage does not stretch like twisted line and, because its construction is more complicated, is more difficult to splice. Also because of the nature of its construction, braided cordage is more useful for certain applications (pulleys and blocks) than others (docking boats).

How carefully cordage is made can suggest its intended end use. A rope whose yarns appear uneven with sloppy or variable construction may indicate unequal tensions during manufacture with a concomitant reduced intended service life. Equal tensioning of yarns, plaits and strands is typically found in quality, high-performance cordage. Tight yarns or plaits will take on a disproportionate amount of tension as the rope is loaded and will prevent premature breakage.

Rope and, more generally, cordage examinations are of interest to the forensic scientist because of the numerous and sometimes particular constructions used for those materials. Manufacturers can often be identified simply through the construction details of cordage. The basic types of cordage construction are shown in Figure 9.8. The term cordage encompasses all types of rope, twine, thread, or any other set of yarns twisted, braided, or entangled to produce a linear construction. A piece of cordage is constructed of two or more strands or plies (singular, ply) which are held together by some form of entanglement, such as twisting or braiding. The number of fibers in each yarn should be counted; this is usually possible with filament fibers but is meaningless with staple fibers. The number of fibers in each yarn should be counted because the number of fibers may vary (with the manufacturers' intent) from yarn to yarn (Figure 9.9).

Some cordage has what is called a core that consists of some number of fibers or yarns in the center of its structure. The fibers or yarns may themselves have some sort of structure that must be dissected. The core may be made of fibers or yarns of all the same color, several of the same color, or they may all be different colors. Often cordage manufacturers use what are called "trash fibers" in the core of their cordage. These are simply filler that do not affect the end use properties of the cordage for the consumer but create bulk so the cordage retains its diameter and shape. Some cores, however, fulfill a specific use and are indicative of the end products design and use. Specialty ropes, for example, used in climbing will have high-strength fibers, such as Spectra®, that are very strong but very

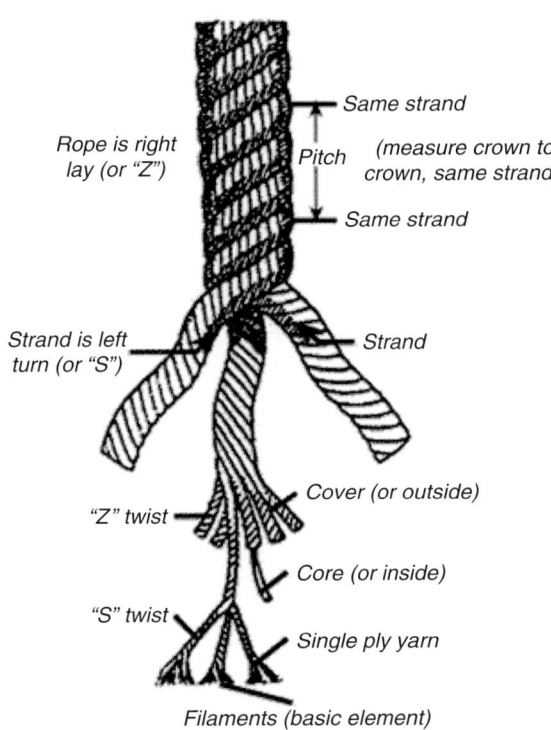

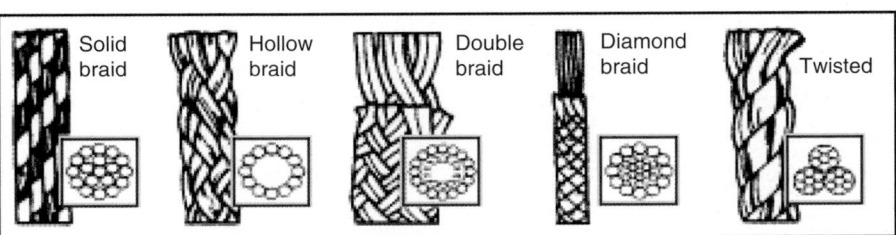

Figure 9.8
Basic rope construction types.

light. These specialty fibers are often expensive and/or rare and provide a guide to the significance of finding that type of cordage at a crime scene.

Another feature of cordage that should be kept in mind is a "tracer." A tracer is some type of identifier intentionally integrated into the construction of the cordage by the manufacturer, such as a thread or yarn of a particular color or variegation. Tracers can also take the form of polymer or paper strips with the company's name or logo printed on them (Figure 9.10). Manufacturers keep relatively quiet about their tracers to avoid copycats and retain the specificity of the identifier. Tracers are part of a quality process that entails product stewardship from "cradle to grave" in all aspects or end products in which the cordage may be used. Liability issues can arise and identification of a product or product

Figure 9.9

Each element in a rope must be examined; the number of fibers per yarn may vary from ply to ply in the core as well as the outer sheath.

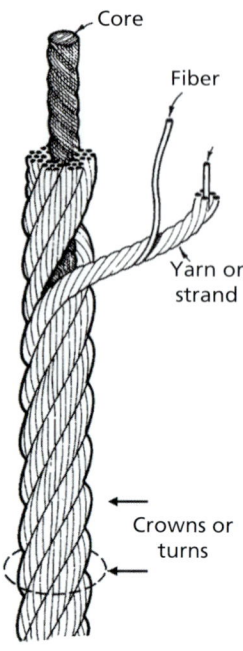

Figure 9.10

Manufacturers may put tracers, colored yarns or printed strips with the company name on them, to identify their products and act as protection against copycats.

component is critical. Tracers are definitive indicators of the manufacturer of a particular item of cordage and can, as we will see, lend immeasurable value to the significance of an item of evidence.

The examination of cordage is an intricate and complicated process; it can, depending upon the construction of the cordage, be quite time consuming. The cordage should be photographed before any analyses are conducted in order to provide documentation of its original condition. Other evidence, such as hair, blood, paint, etc., which may require separate analysis should also be removed. Any physical damage, such as worn areas, cuts, or breaks, should be documented.

The general characteristics of the cordage should be examined and documented, including any severed ends that may physically match other evidentiary cordage, knots and/or ligatures, the dimensions of the cordage, and the color, number, and type of construction components that make up the cordage. The twist or lay, if any, of the strands or plies should be recorded as well as the crowns/turns per inch.

The identification of cordage therefore is done through the determination of its components and construction. Because cordage examinations may be complex and the forensic scientist may not perform them on a daily, weekly, or even monthly basis, a laboratory should develop a checklist for these exams (Wiggins, 1995). The checklist should include, but not be limited to, the following characteristics:

- diameter;
- staple or filament fibers;
- twisted, braided, or unorganized;
- twist direction ("S", "Z", or "0" (zero));
- crowns or turns per inch;
- number of plies or braids;
  - twist of each ply or braid,
  - crowns or turns per inch,
  - number of filaments in each ply or braid;
- core, if any:
  - twist,
  - crowns or turns per inch,
  - number of filaments;
- color(s);
- coatings, if any;
- tracers, if any:
  - coatings,
  - twist,
  - crowns or turns per inch,
  - number of filaments.

The fibers of each construction component should be analyzed with the appropriate microscopic and instrumental techniques available in the laboratory (SWGMAT Forensic Fiber Examination Guidelines, 1999). A description of a rope, for example, may read like:

> Yellow rope, 3′4″ in length, unknotted, dirty with no appreciable wear, both ends cut and frayed (no end match), ½″ diameter, 6 strands, Z-twist. Each strand ⅛″ diameter, 3 plies, S-twist. Each ply 75 filament fibers. No core, no tracer.

It is important not to bring a questioned piece of cordage in contact with a known cordage sample from which it is suspected to have originated (to determine, e.g. if a physical end match exists) until you have performed a preliminary examination of the questioned specimen. When you cut a sample to determine the construction of cordage, do not cut it from the ends if there is a possibility of physically matching a questioned specimen to a known specimen. The known sample should be taken away from the existing ends and the sampled location should be labeled as "known taken."

The black and orange rope samples presented several interesting features.

All of the rope samples were 10 mm in diameter, 3 plies twisted in the Z direction, to a degree of 3 crowns per inch. Each of the plies had four strands, 3 black and 1 orange twisted in the S direction. Each of the strands had 65 off-round olefin filaments, 255 μm in diameter; as the fibers were olefin, they were pigmented. One of the black strands had a tracer, a 1.3 mm by 20 μm wide clear polyester film with "Tubbs Rope Works, Inc." in black print. The ropes had no core.

Interestingly, while the black fibers appeared black grossly and opaque microscopically, the orange fibers were orange to the naked eye but appeared gray in microscopic color. This phenomenon has been observed by the author previously but no explanation could be found in the literature. None of the fibers exhibited fluorescence, which was to be expected because of the pigmentation.

The tracer provided a clear lead to the manufacturer of the rope. Tubbs Rope Works, Inc. was contacted and the following information was offered. The rope itself was a common type of rope that many manufacturers produced, although not all to the same construction specifications found in the evidence ropes. The tracer, however, was another matter. Because it identified the company, the source of manufacturing was known. More important, *that particular style of tracer had not been made by the company since 1988*. It was therefore impossible to go to the local hardware store and purchase "identical" rope: the evidence ropes were nearly unique in the circumstances of the crime.

This small but critical fact is a prime example of the paradox of mass manufacturing. Despite the rope being produced in large amounts, the various parameters of manufacturing methods and standards, quality control methods, general use, time, location, and specific use in the crime in question make it almost "one of a kind." This context was important in conveying the significance of the ropes to the jury, especially because a DNA-like statistical approach was untenable (Houck, 1999). Conceptually, this established what archaeologists call a *terminus post quem*, or "time after which," of 1988, meaning that after that point, the artifact in question does not appear in the archaeological record. A similar idea has been established with so-called "dead fibers," those that are no longer actively manufactured). If these fibers are encountered in casework, the source is known to be at least as old as the last date of manufacture of the fiber and its incorporation into a textile. A forensic *terminus post quem*, which is measured in, at most years, is more exacting than in archaeology, where it is measured in hundreds or thousands of years. Additionally, the date may be well-documented by the manufacturer for product liability or product stewardship reasons. In the absence of applicable statistical methods for trace evidence, conceptual portmanteaus, such as *terminus post quem*, can be successfully applied to aid the trier of fact in understanding the significance of trace evidence.

CONCLUSION

The jury deliberated for 2½h before deciding that Butters was guilty of the second-degree murder of Mahtah. Butters was sentenced to 15 years in prison for his crime.

REFERENCES

Apsell, P. (1981) "What are dyes? What is dyeing?," In *Dyeing Primer*. American Association of Textile Chemists and Colorists.

Brady, G.S. and Clauser, H.R. (1991) *Materials Handbook,* 13th edn. New York: McGraw-Hill Inc.

Hartshorne, A.W. and Laing, D.K. (1991a) "Microspectrofluorimetry of fluorescent dyes and brighteners on single textile fibres: Part 1 – Fluorescence emission spectra," *Forensic Science International*, 51, 203–220.

Hartshorne, A.W. and Laing, D.K. (1991b) "Microspectrofluorimetry of fluorescent dyes and brighteners on single textile fibres: Part 2 – Colour measurements," *Forensic Science International*, 51, 221–237.

Hartshorne, A.W. and Laing, D.K. (1991c) "Microspectrofluorimetry of fluorescent dyes and brighteners on single textile fibres: Part 3 – Fluorescence decay phenomena," *Forensic Science International*, 51, 239–250.

Houck, M.M. (1999) "Statistics and trace evidence: the tyranny of numbers," *Forensic Science Communications*, 1(3) [available on-line at www.fbi.gov].

Houck, M.M., Siegel, J.A. and Thorsen, A.C. (1999) "A large-scale transfer study with fluorescent target fibers," *American Academy of Forensic Sciences*, February.

Park, J. and Shore, J. (1999) "Dye and fibre discoveries of the twentieth century, Parts 1 and 2," *Journal of the Society of Dyers and Colourists*, 157, 207.

Rost, F.W.D. (1992) *Florescence Microscopy*, Vol. I. New York City, NY: Cambridge University Press.

SWGMAT (1999) "Scientific working group for materials, forensic fiber examination guidelines," *Forensic Science Communications*, 1(1) [on-line at www.fbi.gov].

Yeager, J.I. and Teter-Justice, L.K. (2001) *Textiles for Residential and Commercial Interiors*, 2nd edn. New York: Fairchild Publications.

Wiggins, K. (1995) "Recognition, identification and comparison of rope and twine," *Science and Justice*, 35(1), 53–58.

AUTHOR INDEX

Adolf, F.P. 208, 210
Aitken, C.G.G. 24, 26
Aitken, L. 141, 162
Appleyard, H.M. 4, 33, 35, 51
Apsell, P. 237, 359
Arvestad, L. 52
Aspland, J.R. 143, 162

Baldwin-Jedele, K. 134, 163
Biermann, T 209, 210
Biermann, T.W. 102, 134, 162, 163
Bisbing, R. 131, 132, 163
Bloss, F.D. 121
Bock, J.H. 173, 174, 189
Brady, G.S. 239, 249
Brinkmann, B. 54, 62, 88
Brunner, H. 4, 35, 42, 51
Bruschweiler 102, 104
Buckleton, J.S. 25
Budowle, B. 134, 151, 164, 239
Budowle, B. 164
Butler 164
Bürger, H. 54, 59, 86

Caddy, B. 9, 25
Cage, P.E. 42, 52
Cavalli-Sforza, L.L. 135, 164
Charles, E. 190
Chinherende 101, 104
Claiborne, Stephens, J. 47, 52
Clauser, H.R. 239, 249
Coman, B. 4, 33, 35, 51
Conan, Doyle, A. 27, 51
Connelly, R.L. 143, 163
Cook 101, 104, 144
Cook, R. 101, 104, 163
Coquoz, R. 47, 51
Cothern, J.E. 54, 61, 87
Curran, J.M. 25

Dabora, S.L. 135, 164
Dana, E.S. 121
David, V.A. 47, 52
Davis, R.W. 135, 164
Deadman, H. 123
Decke, U. 210
Deer, W.A. 121
Delly, J.G. 122, 173, 190
Devaney, J.M. 134, 163
DiZinno, J.A. 164
Dunlop 104
Dunlop, J. 44, 51
Dwyer, J.D. 189

E.D. 163
Enz, B. 54, 61, 87
Evett, I.W. 42, 52, 157, 163

Few, P.W. 190
Fiddes, F.S. 5, 26
Flint, O. 173, 189
Ford, W.E. 121
Freckleton, I. 51
Fregin, A. 134, 164
Frei-Sulzer, M. 18, 25, 141, 163
Fridez, F. 47, 51
Friedl, W. 134, 163

Gaensslen, R.E. 172, 190
Galvin, C. 190
Gantner, G.E. 170, 189
Garbolino, P. 24, 26
Gaudette, B.D. 44, 45, 51, 133, 137, 139, 154, 163
Girara, J.E. 134, 163
Graves, W.J. 121
Greenish, H.G. 189
Grieve, M. 1, 19, 25, 26, 67, 88, 101, 102, 104, 209, 210
Grieve, M.C. 19, 44, 51, 101, 102, 104, 162, 163

Griffin 101, 104
Gross, M. 134, 163

Haddock, P.S. 44, 51
Hartshorne, A.W. 249
Heinrich, E.W. 121
Hicks, J. 35, 48, 51
Hicks-Champod, T.N. 25
Holinski-Feder, E. 134, 134, 163
Holmberg, A. 47, 52
Hopen, T.J. 105
Houck, M.M. 1, 25, 134, 135, 141, 142, 149, 163, 164, 210, 233, 239, 249
Howie, R.A. 121

Ivaskevicius, V. 134, 164

Jackson 101, 104
Jackson, B.P. 173, 190
Jackson, G. 4, 39, 43, 52
James, D.C. 134, 164
Jenkins, T. 135, 164
Jin, L. 135, 164
Jochem, G. 53, 54, 87
Jozwiak, S. 135, 164
Jungck, M. 134, 163

Keeping 139, 163
Keller, G. 134, 163
Kelly 101, 104
Koff, C.M. 135, 164
Krauá, W. 54, 87
Kuppuswamy, R. 83, 87
Kwiatkowski, D.J. 135, 164

Laing, D.K. 249
Lane, M.A. 189
Lautenbach, L. 59, 87
Lee, H.C. 172, 190
Leitner, T. 47, 52
Lin, A.A. 135, 164

Liu, W. 134, 164
Locard, E. 1, 25, 210
Lohse, P. 134, 163
Lowrie, C.N. 4, 39, 43, 52
Lundeberg, J. 47, 52
Lynch, E. 189
Lyons, L.A. 47, 52

Mange, M.A. 121
Mangold, E. 134, 163
Margot, P. 101, 102, 104, 164
Marino, M.A. 134, 163
Martin, E. 18, 25
Masakowski, S. 54, 61, 87
Maurer, H.F.W. 121
McCrone, L.B. 122
McCrone, W. 12, 26
McCrone, W.C. 121, 122, 173, 190
McPhee, J. 122
Mehdi, S.Q. 135, 164
Meitinger, T. 134, 163
Menotti-Raymond, M. 47, 52
Miller, K.A. 135, 164
Millette, J.R. 187, 190
Moeller, J. 179, 190
Moeslein, G. 134, 163
Moore, J.E. 48, 52
Morse, S.A. 122
Muller, C.R. 134, 164
Muller-Koch, Y. 134, 164
Murken, J. 134, 164
Murray, R.C. 122

Nehse, K. 191
Neubert-Kirfel, D. 205, 210
Norris, D.O. 189

O'Brien, S.J. 47, 52
Oefner, P.J. 135, 164

Oldenburg, J. 134, 164
Oxborough, R.J. 42, 52

Pabst, H. 6, 54, 62, 87
Pagliaro, E.M. 190
Palenik, S. 18, 26
Palenik, S.J. 18, 122, 173, 190
Palmer 89, 101, 104
Park, J. 237, 249
Peabody, A.J. 42, 52
Petraco, N. 19, 26, 122
Piperno, D.R. 182, 190
Plaschke, J. 163
Polanskey 164
Ponnuswamy, P.K. 83, 87
Putnam, B. 6, 54, 63, 87

Rechtzigel, K.J. 134, 164
Replogle, J. 164
Roberts, P.S. 135, 164
Robertson, J. 25, 26, 67, 88
Rochat, S. 47, 51
Rosen, B. 47, 52
Rost, F.W.D. 240, 249
Rost, S. 134, 164
Roux, C. 101, 102, 104, 164
Row, W.F. 54, 61, 87
Rudram, D.A. 157, 164
Ryland, S. 210

SWGMAT 146, 164, 243, 250
Savolainen, P. 47, 52
Schackert, H.K. 134, 163
Schaidt, G. 59, 87
Schiller, W.R. 54, 88
Schneck, W.M. 165
Schwarz, W. 54, 62, 88
Selby, H. 51

Shore, J. 237, 249
Siegel, J. 149, 164
Siegel, J.A. 239, 249
Simon 172, 190
Smith, D.I. 134, 164
Smith, J.K. 134, 163
Smith, S. 5, 26
Snowdon, D.W. 173, 190
Stoecklein, W. 26
Stoiber, R.E. 122
Stoney, D. 141, 162
Stritesky, K. 54, 87

Taroni, F. 24, 26
Teige, K. 54, 62, 88
Tessarolo, A.A. 44, 45, 51
Teter-Justice, L.K. 238, 250
Thibodeau, S.N. 134, 164
Thorsen, A.C. 239, 249
Tridico, S.R. 27
Turner, D. 57, 88

Uhlen, M. 47, 52
Underhill, P.A. 135, 164

Vogelsang, H. 134, 163
Vollrath, D. 135, 164

Weber, B.H.F. 134, 164
Wheeler, B.P. 211
White, K. 134, 164
Wiggins, K. 19, 25, 247, 250
Wildman, A.B. 35, 52
Wilson, C. 144, 163
Wilson, M.R. 164
Winton, A.L. 179, 190
Winton, K.B. 179, 190

Yeager, J.I. 238, 250

Zussman, J. 121

SUBJECT INDEX

1:1 taping
  collecting trace evidence
    DNA analysis 193
    metal roller device 194
    tape 194
    tiny smear of blood 192
  interrogation 195
  reconstruct a source
    blood 191, 192, 194
    DNA analysis 193
    fingerprints 192, 194
    footprints 192
    toothpicks 192

amylopectin
  commercial sources
    cereals 176, 178, 179
    root tubers 176
  gelatinization
    Maltese cross birefringence pattern 178
  insoluble granules 175
  "Maltese cross" extinction pattern 176
animal hairs
  cat hairs
    indistinct fibrillar root 48
  dog hairs
    spade shaped roots 35, 48
  identification
    root shape 39, 48
Anthrone micro-color test 180
aramid fibers 7
automotive coating 6, 15

barrier
  analysis
    carpet fibers and color 237
    finished dye 237
    visual comparison 237
  crime scene
    "Air Canada" tape 236
    carpet-type fibers 236, 242
    fingerprints 234, 235, 236
    saliva samples 235

resin-solvent paint base
  beta radiation 239
  fluorophores 239, 240
  half-life 239
ropes and their construction
  high-strength fibers 244
  "trash fibers" 244
Bayes' theorem
  degree of subjectivism 24
  objective probabilities 24
blood 28, 42, 54, 108, 134, 161
blood samples
  in head hairs 152, 156
  qualitative (non-numeric) estimates 156
blood smear 193
Bureau of Alcohol, Tobacco and Firearms
  physical evidence
    cereal murder 165
    garbage cans 166
    life insurance policy 166

carpeting samples 223
  crushed fibers 223
  fiber additives 219, 224
  FTIR spectroscopy 224
    hair fragments 220, 228
    re-mounting the samples 228
  man-made fibers 224
  navy carpet 223
  purple cotton 223, 226, 229
  purple dye 223, 226
  purple fibers 220, 221, 223
cat hairs
  DNA typing results
cereal murder
  Collection of physical evidence 165
chromosomes 134
chyme 170
coagulation 172
collecting reference material
  advantage
    crime-related areas 204
    direct link 204

collecting reference material (*contd*)
  advantage (*contd*)
    fiber type
    fiber-contaminated areas 204
  analysis
    black polyester fibers 201
    black rayon fibers 201, 203, 204
    black viscose fibers 201
    brown-black cotton fibers 201
    crime scene 199, 200, 201
    direct fiber transfer 199, 209
    gray acrylic fibers 201
    indirect transfer 199
  fiber environment 198
  fiber "pill"
    DNA 203
    tiny bloodstain 203
  fiber type
    application of dye 208
    color 208
    shade 208, 209
    single fiber type 206, 208, 209
colored manufactured fiber
  acrylic 143, 144, 153, 154
  nylon 143
  polyester 143
  two-way transfer 144
combination 208
common dye
  indigo 217
common fiber types
  blue cotton fibers 71, 75, 76, 142, 156, 201
  Caucasian head hairs 150, 151
  off-white cotton 142
  white cotton 58, 142, 196, 217, 221, 225, 226, 229
contamination 94, 136
cosmetics
  glitter particles 2
  lipstick 2
  makeup 2
  polished fingernail 2, 3
  significance 100
crime scene personnel
  bulk evidence collection 9
  crime scene samples 9
  packaging 9, 20, 229
criminal justice system 125
cross-examination testimony
  rebuttal by the prosecution 230

density separation method
  feldspar grains 113
  polarized light microscopy (PLM) 114
  quartz 113, 114, 115
dichotomous plant cell keys
  food standards 175
  slide sets 175
dichroic mirror 242
discriminating test procedures 125
DNA analysis 19, 126, 130, 132, 139, 171, 193
DNA testing
  DNA databases 126
  forensic DNA analysis 126
  sexual assault 126
dog hair 33, 48
dye application 201, 209
dye peaks 18

evidence 124–125
  criminal case 124
  erroneous convictions 123, 125, 132
  homicide 124, 128
  prosecutor's evidence 124
evidence ropes 248
evidential value
  assessment
    live trials 17
    population studies 16, 101, 102
    studies on blocks of color 17
    target fiber studies 16, 101, 141, 142, 144, 154
excitation 219, 224, 239, 240, 241, 242
expert witness 53, 124, 126, 127, 128, 129, 227, 231
eyewitness 27, 46, 81, 128, 215, 216

FBI Laboratory
  hair match 133, 137, 138, 150
Federal German Supreme Court
  DNA result 147, 207, 208
  fiber result 207
  first fiber transfer 209
fiber investigation 196
  direct fibers 197, 199, 204, 209
  fiber population 195, 198, 199, 201, 204
  indirect fibers 197, 199, 204, 209
  rayon fibers 196, 197, 198, 199, 200, 201, 203, 204
  see also original clothes 197, 199, 204, 209
  UV/Vis 196, 208
fiber matches
  man-made fiber
    debris fibers 147, 148, 154
    tire iron 154
    white powder 154
fiber-plastic fusions (FPFs) 53, 54
  leather varnish 60, 61
  PC-crash 57
  pinched fibers 62, 63
  plastic coating marks
    local melting 55
    textile printings 60

filter types
    conventional glass filters 241
    filter cubes 241
    longwave pass band filters 241
    pass band filters 241
    shortwave pass band filters 241
floor mat 221, 223, 224, 225, 227, 228, 229, 230
foreign fibers
    FTIR spectroscopy
        "fingerprint" pattern 221
        polymeric smear 221, 225
forensic evidence
    DNA profile 28, 42
    single bloodspot 27, 42
forensic fiber analysis
    central concern 136
    contamination 136, 147, 148, 149
    polymer class 135, 196, 219
    type of textile material 135
forensic hair analysis
    human hair comparison
        color variation 130
        growth cycles 130
        hair growth 131
        loss of pigmentation 131
        medullary indices 36, 37, 39, 42, 48
        three-dimensional 131
forensic scientist 125–126
Fourier transform infrared (FTIR)
    non-destructive identification 219
    polymer comparisons
        fibers 219, 220, 221
        paint 219, 239
        rubbers 219, 221
        thermoplastics 219, 221

garbage cans
    analysis of the white droplets 184
    gypsum-containing wall texture particles
        clay 184
        diatomaceous earth 184
        titanium dioxide 184
    hand swabbings 185
    paint overspray 184
gastric fluid stains
    see also vomit 171, 172, 174, 181
gastric juices 170, 172
genetic fingerprinting
    DNA typing 19, 47, 126, 135, 136, 139, 155
German Federal Police Laboratory
    fiber recovery 17

hair
    chromosomes 134
    DNA 134
    nuclear DNA 134

hair and fiber analysis
    evidence recovery 96, 129, 148
    screening 129, 148
hair and fiber defense expert 156
hair and fiber evidence
    human hair 129, 130, 132, 135, 137, 139, 151, 162
    textile fibers 140–145
hair match 133, 137, 138, 150
hair of the dog
    analysis
        animal hairs 31
        tape-lifts 40
    Australian Cattle dog
        colorings 38
        dog's bedding 41
        tape-lifts 40
    crime scene
        Australian Cattle dog 28, 29
        veterinary examination 29
    investigative process
        animal hairs 30, 31
        designer label clothes 30
        veterinary examination 29
    primary transfer 39, 43
    secondary transfer 39, 43
    trial 44
hair types
    guard hairs
        primary 33
        secondary 33
    overhairs 33
    underhairs 34
hair-training programs 132
head hair samples 150, 151, 215, 216
human hair
    damaged hairs
        criminal activity 137
    elemental data 230
    environmental factors 137
    evidential value 134, 136, 137
    forensic hair comparisons 133, 134, 139
    hair match 133, 137, 138, 150
    hair proficiency tests 139
    hair samples
        head and pubic hairs 135, 138
    microscopic characteristics 137, 138
    natural characteristics 137
    scale patterns 14, 35, 48, 230
    two-way transfer 138, 144
hydrochloric acid 170

identical fiber
    composition 102
    construction 101
    contamination 136, 147, 148
"identical" rope 248

importance of trace evidence
  actual hair and fiber results
    FBI Laboratory 149
    location 150
  appeals
    Habeas petition 159
    Magistrate and District Court 160
    South Dakota Supreme 158, 159
    United States Court of Appeals 160
    United States District Court 159
  black rayon fibers 152
  case
    radio-tracking device 145
  FBI Laboratory
    hair samples 146, 150
    Oldsmobile trunk 146, 151
    trunk liner sample 146
  General Motors
    automobiles 152
  laboratory approach
    DNA 147
    fingerprints 147
    mounting medium 148
  optical properties
    probative fibers 149
  South Dakota Forensic Laboratory
    mounting medium 146, 148
    mounting of hairs 146
  verdict 158, 159
indigo 79, 81, 142, 217

Kentucky State Police Forensic Laboratory 211, 215

laboratory examination
  acrylic fiber 92, 98
  cat hairs 92, 94, 95, 98
  dye components 97
  Fourier transform infrared spectroscopy (FTIR), 97, 196, 219, 221, 224, 225
  microscope slide 97
  microspectrophotometry (MSP) 97
  thin layer chromatography (TLC) 97
liability issues
  identification of cordage 247

mammalian hairs
  medulla 36, 48
man-made fibers 83, 144, 224
map
  Alabama along I-85 109, 110
  Crowhop, Georgia area 109
  Georgia border 109, 110
  soil profile map 109, 111
micro-FTIR
  co-polymers 5
microscopist 174

mitochondrial DNA analysis
  hair shaft 134
mitochondrial DNA match 135
mute witness 128

negative finding
  absence of evidence 3

occupant kinematics 229
optical microscopy
  complementary filters 240
  principles 240
original clothes 197, 199, 204, 209

paint technology 8
pepsin 170, 172
petrographic similarities 182
phytoliths
  inorganic 182, 184, 187
  organic 182, 187
pigments
  organic and inorganic 8
plate-like pigments
  analysis 15
  refracting materials 8
  three-dimensional 8
PLM examination 114
polarized light microscopy 10, 114, 121, 171, 224, 242
  food particles 171, 173, 175
  vomit stains 168, 170, 171, 172, 173, 174, 179, 181
pollen grains 2, 13, 105, 108
polymer
  amylopectin 175
  amylose 175, 177
  polyoxymethylene polymer 84
  starch 175
polymer types 67, 143, 225, 229
polymeric smears 221, 225
polymorphs 9
polypropylene 7, 10, 75, 153, 154
Ponderosa pine needles 182
PyGC-MS
  paint additives 15
pyrolysis gas chromatography (PGC) 219

race estimation
  Caucasoid 135
  Mongoloid 135
  Negroid 135
Raman spectroscopy
  automotive coatings 6, 15
red acrylic fibers 92, 144
reddish soil 107, 108, 109, 113, 118, 119
renin 170, 172

rope construction
  clockwise 243
    see also S-laid or S-twist
  counterclockwise 243
    see also Z-laid or Z-twist 243
rubber sole
  soft polymer 221

safety glass 220
saliva 170
saliva stains 134
scanning electron microscope (SEM) 229
SEM photographs 230
semen 134
shoe soles 221, 222, 225, 229
shoe tread patterns 220
S-laid or S-twist 243
  see also clockwise
soil 18, 105–121, 165, 167, 181, 182, 183, 185
soil examinations
  color and texture 109
  stereomicroscopical examination 109
soil makeup
  addition of man-made contaminants 105
  indigenous additions
    pollen grains 105, 108
  mixing of two different soils 105
  natural amendments
    flower bed 105
    lawn 105
    sand 105
  natural weathering 105
  physical alteration 105
soil particles 182
soot
  carbon soot particles
    aciniform 187
    cellulosic material 187
    cenospheres 187
    char fragments 187
    microgel 187
  file cabinet 165, 187
  soot particles 170, 187, 188
South Australian Dog Breeder's Association 38, 43, 51
"spider-web" pattern
starch
  carbohydrate storage product 175
    chlorophyll containing plants 175
stereobinocular microscope
  food particles 171, 173, 175
  food products
    aqueous iodine solutions 174, 177
    Oil Red "O" 174, 181
    Safranin 174
    Toluidine Blue 174
    Trypan Blue 174, 175, 178, 181

  identification of food 172, 173, 174, 175
  vomit stains 168, 170, 171, 172, 173, 174, 179, 181

target fiber studies 16, 101, 141, 142, 144, 154
ter-polymers
  chemicals 5, 9, 10, 12
terrorist attacks
  World Trade Center 135
testing for gastric enzymes
  feces 172
  saliva 172
  urine 172
  vomit stains
    coagulation 172
  whole blood 172
textile fibers
  laboratory results 129, 140, 226, 228
  testimony 129, 140
textile industries 6
thermoplastic 54, 55, 58, 61, 62, 69, 79, 83, 217, 219, 221
trace element determination 15
trace evidence
  1:1 taping 2, 192, 194, 195, 204, 207, 209
  car examination
    chains/padlocks 100
    debris from the car 99
    surface debris 96, 97
    taping 96, 97
    vacuuming 96, 146, 150
  cross-transferred 1
  direct sunlight 63
  elemental analysis 13, 22
    fiber examination 13, 14, 16
    instrumental techniques 13, 22
  eye witness testimony 53, 127
  fiber fragments 63
  G.I.F.T. principles 46
  house examination
    cat hairs 94, 95
  human hairs, fibers, paint, and glass 2, 127
  initial laboratory examination
    nature of the gag 92
  intelligence-led investigation
    fingerprints 90
    identity of the deceased 90
    postmortem examination 89
  knitted clothing 94
  light 63
  outerwear 63, 73
  police investigation
    mobile phone records 94
  quality assurance 20, 21, 23, 27
  recent advances
  results of laboratory examination 97

trace evidence (contd)
  seating position 63
  serological analysis
    bloodstains 54, 108, 203
    blood type 108
    pollen grains 105, 108
    semen 134
  trace evidence analyst *see* forensic scientist 1, 9, 10, 12, 19, 22
  wet surface 63
trace evidence unit 211, 215, 216, 223, 225
trace material
  examination
    Fourier transform infrared 67, 97
    microscopy 51, 67
    microspectrophotometry 67
    transferred fibers 43, 55, 67, 73
  hairs
  molten plastic 61, 79
  thermoplastic part 61
tracer 245, 246, 247, 248
traces of vomit
  "fast food" hamburger meal 171
traffic accident investigation
  examination of the bloodstains 54
  head-on collision 54
  Mercedes
    Windshield 214, 215, 216, 219
traffic accidents
  case example 1
    ABS 72
    embedded blue denim cotton fibers 70
    polyvinylchloride material 71
    seating positions 70, 72
  case example 2
    embedded blue denim cotton fibers 73, 74
    sweatshirt 75
    VW transporter 73, 74
  case example 3
    clothing 75, 77
    cotton fibers 75, 76, 77
    examination of the wreck 75
    fusion mark 75, 77
    melting marks 75
    PVC material 75
    steering column 75, 76
  case example 4
    denim cotton fibers 79, 80
    fusion marks 80
    transferred plastic materials 81
    VW Golf 78
  case example 5
    hit-and-run cases 82
    police check-point 81

  case example 6
    blue-green polyamide 82
    bomber jacket 82
    denim fabric 82
    drunken pedestrian 82
    fiber case 83
    textile pattern 82, 83
  case example 7
    bicycle rider 84
    car: BMW 84, 85
    cross-transfer of trace material 84
    fusion marks 84
    polyoxymethylene polymer 84
    windscreen washer 84
  seating arrangement
    reconstruction 70, 77, 84, 85, 86
twill-weave pattern
  denim 225
  dungaree 225
  gabardine 225
  houndstooth 225
  jean 225, 229
  melton 225
  whipcord 225
two-lane roadway 211
type II error 132
uncommon fiber types
  Monsanto Chemical Company
    carpet fiber 143, 152

UV/Vis 196, 208

value of soil evidence
  case 1
    burglary 106, 107
    clumps of soil 106
    dirt floor 106
  case 2
    rape suspect 106
    trilobal acetate fibers 106
  case 3
    foraminifer 107
    trucking company 107
  case 4
    court-ordered restitution 107
    mineral matter 107
    plant fragments 105, 107
  forensic laboratories 105, 121, 132
vehicle dynamics 214, 215, 227, 229
vehicular accidents
  accident reconstruction 216
  pedestrian–vehicle investigation
    see also traffic accidents
  analysis of the trace evidence 216
visual comparison 237
vomit 166, 169–171, 172

"who do you believe?"
  analysis
    Kentucky State Police Forensic
      Laboratory 211, 215
    Trace Evidence Unit 211, 215, 216, 223, 225
  court proceedings
    re-examination 227
  crime scene
    acute peritoneal hemorrhage 213
    blunt force trauma 213
    cardiorespiratory insufficiency 212
  investigation
    "positions" of the vehicle 214

wool fiber 144

X-rays 15, 22, 239

Z-laid or Z-twist 243
  see also counterclockwise